The Open University

Arts/Mathematics/Science/Technology
An Inter-faculty Second Level Course in the History of Science

Science and Belief: from Copernicus to Darwin

Block VI Units 15–16

The New Outlook for Science

The Open University Press

COVER ILLUSTRATION
Faraday in his laboratory at the Royal Institution, painted by Harriet Moore in 1852 (Royal Institution of Great Britain).

The Open University Press,
Walton Hall, Milton Keynes.

First published 1974.

Designed by the Media Development Group of the Open University.

Printed in Great Britain by
MARTIN CADBURY, A SPECIALIZED DIVISION OF SANTYPE INTERNATIONAL, WORCESTER AND LONDON

ISBN 0 335 00786 4

This text forms part of an Open University course. The complete list of units in the course appears at the end of this text.

For general availability of supporting material referred to in this text, please write to the Director of Marketing, The Open University, P.O. Box 81, Walton Hall, Milton Keynes, MK7 6AT.

Further information on Open University courses may be obtained from the Admissions Office, The Open University, P.O. Box 48, Walton Hall, Milton Keynes, MK7 6AB.

1.1

Contents

Set Books

C. C. Gillispie, *Genesis and Geology*, Harper and Row.

D. C. Goodman (ed.), *Science and Religious Belief: A Selection of Primary Sources*, John Wright and Sons/The Open University Press (ANTHOLOGY I).

R. Hooykaas, *Religion and the Rise of Modern Science*, Scottish Academic Press/Chatto and Windus.

C. A. Russell (ed.), *Science and Religious Belief: A Selection of Recent Historical Studies*, University of London Press/The Open University Press (ANTHOLOGY II).

Unit 15
Nature and History

Part One prepared by Professor R. Hooykaas for the Course Team
Part Two prepared by Clive Lawless for the Course Team

——— Who can paint
Like Nature? Can Imagination boast,

ILLUSTRATION OVERLEAF: *'Flower Garden' from James Harvey,* Meditations and Contemplations, *volume 1, 1796 (courtesy of the Evangelical Library).*

Contents

Introduction (by Clive Lawless)

The arrangement of this unit follows the general pattern of Units 6 and 11; this Introduction is followed by Part One, an essay written by Professor Hooykaas. In the margin are references to questions which will be found in Part Two. It is suggested that you should complete the question below before starting on Part One. When you come to Part One take each section separately and complete the appropriate question(s) in Part Two as they occur, before starting on the next section of Part One. In studying this unit you will be asked, in the questions in Part Two, to compare the views put forward in this unit with those put forward in other units, notably Unit 14. Units 14 and 15 have been written by two very distinguished scholars in their own fields, Dr Alan Richardson and Professor Hooykaas, and they have been written quite separately. No attempt has been made to produce a synthesis of their arguments (that is your task!) on what are still live issues in the history of science and its relationship to religious belief.

The ability and readiness to apply critical judgement to all sources, primary and secondary, and to the course units themselves is an integral part of a university course, and it is particularly necessary when dealing with material as complex as that contained in this unit. The fact that material is contained in a unit or in a set book does *not* give it a seal of approval. A set text may be chosen because it puts forward a point of view so that you, the student, can examine it critically in the light of other materials provided in the course and, quite possibly, reject the point of view. You will have noticed, for example, in Unit 11 that certain aspects of *Gillispie* were subjected to critical judgement in this way.

Objectives

After working through this unit[1] you should be able to:

General objective

Assess critically the different views of history presented in Part One and the relationship of these views to the concept of the history of nature.

Detailed objectives

1a Distinguish between the principles governing the study of the history of nature and the history of mankind;

1b Relate the 'historicization of nature' to the Hebrew idea of history and discuss the implications of extending the history of nature into the history of mankind;

2a Identify the different meanings of 'uniformitarianism' and distinguish between the use of the term for a system and for a method;

2b Compare certain specific nineteenth-century approaches to the position of mankind in creation;

2c Analyse the reasons for the modification of uniformitarianism in nineteenth-century geology;

3 Identify the hazards in a discussion about the repetition of events in history;

4a Compare and contrast the approaches to the history of evolutionary thought by Dr Alan Richardson in Unit 14 and Professor Hooykaas in Part One of this unit;

4b Discuss the implications of the idea of progress, and of its transfer from biology to other fields and analyse the results.

[1] There is a certain amount of common ground between these objectives and those for other units.

Preliminary Question

In the introduction to Unit 14 Dr Alan Richardson poses the question:

> Why did the concept about the evolution of the world, of man and of his ideas not become clear until the nineteenth century A.D., and then only within a comparatively small section of 'western' civilization?

Briefly note down your answer to this question and compare this with the answers Richardson gives. (Reference will be made to this question in Part Two.)

Part One (by Professor R. Hooykaas)

1 Methods of Science and History

1.1 The problem

In some popular works on science and on history the history of mankind is considered as just a continuation of the history of nature. In a continuous development from atoms to celestial bodies, the earth, living beings and human culture are depicted in vivid colours, and sometimes a prophetic spirit seems to urge the author to add even the future of the universe and the development of man into superman.

Many historians deem such a view too ambitious. J. Huizinga (1929) spoke about the 'clever, unshaken evolutionists, for whom the history of the world has no longer any riddle', and under whose hands 'the process of the world becomes the simplest possible matter': these are the people who unhesitatingly tackle the history of the whole of mankind, and who are not even satisfied with the human past, but 'in order to get the right perspective, the history of the planet, and of life on earth has to come first.'

Other historians emphasize the intrinsic difference between the history of the creators of culture and the sequence of events the earth and its non-human inhabitants went through. In this unit, moreover, we shall see that the methods of historiography of man and that of nature are essentially different.

Consequently, it seems expedient to consider how far there is or can be a *history* of nature, and which are the analogies and differences between the history of nature and that of mankind. As our images of nature and of history are built up by science and by historiography respectively, we have also to consider the methods of these two disciplines.

Apologies are due for the fact that the problems will sometimes be touched too lightly, but a more elaborate treatment would involve us too much in philosophical issues for which we send the reader back to the recommended literature.

1.2 Science

The typical and fundamental science of nature is physics, and not geology or any other 'historical' discipline. The physicist tries to explain phenomena by reducing them to the interplay of fundamental units (molecules, atoms, electrons, etc.), i.e. by reducing change to identity. The same 'matter' is present in simple bodies and in the thousands of their combinations; the same 'energy' is present in heat, electricity and the mechanical labour performed by them. Identity is sought for, not only in the *being* of things but also in their *change;* from identical situations A ensue the identical consequences B. This repetition of events observed in nature, or elicited by artificial experiments, is expressed in 'laws of nature'. In principle, physical events are repeatable. (Of course, in real life no experiment can be perfectly reproduced; there are always circumstances interfering with the ideal event. The ideal and 'typical' experiment is the 'thought experiment', imagined as being without disturbing influences.)

Physics is most of all interested in what is general, conformable to a 'law' or rule, non-individual. Individuality is of little importance: for electrons the notion cannot be used, with crystals it is difficult to handle. Even in the biological sciences it is unimportant; zoological systematics is interested in the species, the collectivity of individuals.

The situation is different in geography, which describes not only continents, mountains and oceans in general, but also Europe, the Alps, the North Sea in particular, or in astronomy, which not only gives general verdicts about suns, planets and comets, but makes a catalogue of them. Similarly, geology tries to find 'laws' for the rise of mountains in general, but it may also enter into the particular circumstances giving rise to the Alps. Moreover, the sciences dealing with the genesis of the universe and the earth's crust, and with the origin of living beings, are confronted with a never-repeated sequence of events, which is unique and individual. Nevertheless, even then palaeontology and biology will look for general laws within this unique series of events.

1.3 History of mankind

The history of mankind, in which irreplaceable personalities (Henry VIII, Bonaparte, etc.) play an important role, cannot be studied with the methods borrowed from natural science. Here 'understanding' in the sense in which the scientist understands phenomena is impossible, but we seek another kind of knowledge, that of sympathetic understanding, by which the historian relives in his imagination the individual experiences, feelings and thoughts of the people of the past: the poets, artists, politicians, etc., and thus makes the past alive again (Windelband, 1894; H. Rickert, 1929). This re-enactment has to go together with a critical analysis of the motives of individual and collective actions, with a confrontation of arguments and a consideration of analogies and relations between facts past and present (Huizinga, 1929). It goes without saying that natural disasters (inundations, earthquakes, epidemics), and other exterior circumstances influencing human life, cannot be left out. Such a re-enactment[1] is inseparable from ethical and political valuation.[2]

Uniqueness,[3] then, is a salient characteristic of history. Certain patterns of circumstances and events may be repeated, but this is not an identical reproduction. A historical event changes the conditions in such a way that those leading to that event cannot occur a second time.

Not only is the historical situation objectively changed by the course of events, but mankind has a memory of past events and this co-determines human activities. The adherents of the Bourbons in 1815 could try to restore the pre-revolutionary situation, but they could not (and would not) wipe out the recollection of the Revolution. Their actions were determined not only by pro-royalist but also by anti-jacobin and anti-bonapartist feelings. The thoughts of the men who are for a large part responsible for future events are *changed* so much by past events that those earlier events cannot be repeated. The repetition at will, so characteristic of the scientific experiment in physics, is not possible in history. **Question 1**

[1] *Nach-erleben.*

[2] *individualistisch und wert-beziehend.*

[3] *Einmaligkeit.*

Moreover, human beings possess not only individuality but also 'personality': they try to determine their own history and that of their natural environment, whereas the objects of science (stars, the earth, plants, animals, including man as an object of morphological and physiological research) just undergo events. As Pascal said: 'The eternal silence of these infinite spaces frightens me . . . By space the universe comprehends and swallows me as a point; by thought, I comprehend it'.

Figure 1 Blaise Pascal (Mansell Collection).

Economic and social historians, in particular, however, tend to look for laws of the same character as those propounded by scientists. These disciplines are concerned with mass phenomena, with nations and social groups, and their economic and social conditions. Individuals (kings, politicians, generals) play here a much less important role than in political history. Consequently, it is possible to find regularities which, because of a certain degree of similarity in the spiritual and physical conditions of different people, are easily believed to bear a 'scientific' character. As in physics, statistical methods are applied, which eliminate the peculiarities of each individual and produce the 'average' member of the group, just as in a gaseous body an average velocity is attributed to the molecules and even the percentage of 'deviations' from this average may be determined. One knows the percentage of alcoholics, criminals, suicidal attempts per annum within a population, as one knows what percentage of molecules will exceed a certain velocity. This approach to the social sciences received a strong impetus from the Belgian physicist and astronomer Adolphe Quételet (1796–1874).

Figure 2 John D. Rockefeller (Bettmann Archive).

There is, however, one great difference between *social* and *physical* statistical rules: the number of criminals may decrease as a result of a change in economic and social conditions, and people's behaviour may be altered precisely because the outcome of statistical investigations is publicized. Besides, scientific beliefs may influence social behaviour; and conversely, social beliefs may further the acceptance of scientific tenets. 'Social Darwinism' transferred natural selection (which, according to Darwin, 'picks out with unerring skill the best varieties') to industrial competition: 'The growth of a large business is merely a survival of the fittest. . . . This is not an evil tendency in business. It is merely the working-out of a law of nature and a law of God' (John D. Rockefeller; quoted from Hofstadter).

The efforts to apply the methods of natural sciences in human historiography are inspired by a naturalistic conception of history. If one believes that nature and history are perfectly continuous, one feels there is every reason to use analogous methods when investigating them. Different conceptions then are possible:

1 It may be held that events in nature are non-repeatable; the scientific law of change is to be sought for both nature and history. Historical development is embedded in a semi-historical naturalistic system. This is the way of the evolutionistic world view which implies a 'historicization' of nature (when admitting a unique, non-repeatable development) as well as a 'naturalization' of history (when human history is subjected to the same law of evolution as sub-human nature). Or it may be said that:

2 Both nature and history show a perpetual repetition of the same pattern of events. With the Greeks there was a tendency to such a conception. Their natural philosophers tried to discover immutable Being (see *Hooykaas* (SET BOOK), pp. 2–6). Aristotle's system allows for a continual generation and corruption of individuals (except for heavenly bodies, which are permanent and liable to change of place only); but the *species* remain for ever the same. In biology such a conception means that coming-to-be and decline are fully recognized within the life-span of an individual, but that this sequence is expected to be repeated again and again within the same species. In history it implies that a chronological sequence of different particular situations will repeat itself under the influence of the repetitive circle of heavenly constellations. When the whole of the heavens has reached again a certain constellation, nature as well as the history of mankind will have gone through a complete cycle, and not only the subsequent constellation of the planets, but also the events of human history, will be a repetition of the previous cycle.

2 History of Nature

Question 2A
Question 2B
Question 2C

In the eighteenth century scientists began to take more interest in the genesis of nature. The cosmological speculations of Burnet and others, however, met in general with little approval (see Unit 11). Most geologists restricted their investigations to the recent vicissitudes of the earth, and, in order to explain these they refused to posit revolutions affecting the whole earth and having no counterpart in present events.

The rise of historical geology was soon followed by that of the history of life on earth. Towards the end of the century Kant stated that the science of Being had progressed so much that it was now possible to study, besides the old 'natural history' (description of the furniture of inorganic and organic nature), the history of its Coming-to-be.

The dilemma was inescapable: has nature a history, i.e. does it present a sequence of unique, non-repeatable events and situations, *or* is all change in nature a perpetual repetition, like the sequence of generations of plants and animals? This gave rise to two opposing kinds of systems: *uniformitarian* and *non-uniformitarian*.

2.1 Uniformitarianism

Uniformitarianism holds that the same processes of erosion, mountain formation, etc., have always taken place, so that, as far as we can see, the average situation of the earth's crust has remained the same. If there was once a different initial situation, it has left no traces: 'the result, therefore, of this physical inquiry is that we find no vestige of a beginning—no prospect of an end' (Hutton, 1785; 1795). As far as geological research can reach, Hutton held (like Lyell after him) that the thing that is, is the thing that has been, and shall be. This is not in conflict with efforts to find the changes of the past; there *is* a history of the earth, but it goes in circles. Continents arise and disappear, so that the global situation remains the same: 'This world is thus destroyed in one part, but it is renewed in another. . . . This earth, like the body of an animal, is wasted at the same time that it is repaired. It has a state of growth and augmentation; it has another state, which is that of diminution and decay'; 'this world . . . may it not be also considered as an organized body? such as has a constitution in which the necessary decay of the machine is naturally repaired, in the exertion of those productive powers by which it has been formed' (Hutton, 1795). Similarly, Hutton did not assume changes in the organic species. His concessions to 'change', then, do not seem to be greater than those made by Aristotle.

This steady-state system is essentially a-historical and, consequently, typically 'scientific'. Just as physical phenomena are repeated according to fixed laws and physical experiments may be repeated at will, geological phenomena are repeated in a world whose global state remains the same.

Charles Lyell (1830) carried this 'actualistic'[1] method to an extreme, explaining the former changes in the earth's surface not only 'by reference to causes now in operation', but also by attributing to these causes the same energy as they have today (*the methodological principle of uniformity*). This was said not only of the primary, physical causes, but also of the secondary, geological causes. With him, too, a steady-state system of geology was the inevitable result; with him, too, (to use Hutton's words) 'the Theory now proposed explains this operation by known causes . . . powers the most constant, the most general, that act upon the surface of the world'. With Hutton and Lyell the steady-state system and the rigorous application of the principle of uniformity implied that geological events needed a long time and happened gradually. Question 3

[1] See p. 17 below.

2.2 Non-uniformitarian systems

These presuppose that throughout the ages the state of the earth's crust and of the organic world has changed irreversibly. Consequently, in the past and the present not only causes of the same kind, but also causes of a different kind from those active now, must have been in operation, and (or) the energy of causes operating in the past may have differed considerably from the energy shown by these same causes at present.

Those who postulated a general decline of the earth's temperature held that geological change became slower and less violent in the course of time (Hooke, Elie de Beaumont).

A particular form of non-uniformitarian system is *catastrophism*, which holds that in the past there were geological upheavals of such a kind and such a violence as do not occur in modern circumstances. As Cuvier (1826) put it: 'The thread of operations is here broken; the march of Nature is changed; and none of the agents which she now employs would have been sufficient for the production of her ancient works.'

Catastrophism possesses a distinctly *historical* character; it admits unique, non-repetitive events and situations. Whereas the uniformitarian Hutton chose as a model of the earth's history the adult individual, who is maintained by a monotonous repetition of destructive and reconstructive physiological processes, the catastrophist L. Elie de Beaumont (1846) compared the history of the earth's crust with the irreversible development of the individual from early youth to old age.

Catastrophism found a strong support in palaeontology, which seemed to show a rather sudden appearance of higher animal forms in younger strata. This difference between ancient and more recent fauna implies a *history* of the organic world, and—for those who held that biological changes were caused by geological changes—also a *history* of the earth itself. The life of the first inhabitants of the earth 'began . . . under the rule of influences which differ from those that characterize the economy of nature at present' (Geoffroy St Hilaire, 1831).

The British school of 'progressionism' (Buckland, Sedgwick, Hugh Miller) stressed the progress, the sudden rise of higher and more complicated animal forms after major catastrophes, but the latter of course were denied by the uniformitarian Lyell, who adhered to a belief in a steady state not only of the earth but also of the animal kingdom.

2.3 Human history and history of nature

1 It is understandable that Cuvier, who admitted rapid and great revolutions in nature, regarded *human* history as relatively a-historical. He pointed out that for the recent epoch of geology (that since the last catastrophe) it is possible to find out what happened on the assumption that the geological causes are the same (in kind and degree) as those which are still acting today, 'in the same manner as it is found easy to explain past events in political history by an acquaintance with the passions and intrigues of the present day'.

2 Lyell, on the other hand, who regarded events in nature as rather slow and uniform, was struck by the mobility of human history. He emphasized the difference between them. The rise of man as a moral being and a creator of culture was to him a breach of the uniformity, an 'innovation', 'differing in *kind* and *energy* from any before in operation'. He energetically rejected the accusation of having taught *absolute uniformity* throughout all ages; such a doctrine, he says, is fitting for some Egyptian and Greek sects, who too closely connected natural events with moral events, subjugating moral events to the influence of periodically repeated events in heaven.

The moral order, then, is for Lyell a completely new phenomenon, superposed on the physical order without diminishing the latter's uniformity; it is an order in which there is not, as in physics, geology and biology, an endless repetition, but one in which history in the proper sense begins. Even after he accepted Darwin's theory of evolution (that is, after starting to think more 'historically' in biology too), Lyell considered the coming of *homo sapiens* as an interruption in the course of evolution, because with Man a wholly new order of phenomena made its appearance.

3 It might be said that the progressionist Hugh Miller merged the history of nature with that of mankind. He expounded what he considered the first (cosmological and geological) phases of the great divine drama, the general outline of which had been revealed from Genesis to Revelation. Beginning with the creation, in stages, of inorganic nature, followed by that of plants and ever higher animals, and crowned with the creation of Man, the drama continues in the history of mankind, which will find its culmination in a new heaven and a new earth.

It seems strange that a scholar who so much stressed the special position and character of man in the whole of creation should unite these events in *one* picture. But, first, Miller does not admit continuity; secondly, even the quasi-history of geological and biological progress as depicted by him is, like the history of mankind, an '*inside*' historiography. All of it is to him the history of the thought and will of a Person, namely God, the Creator of heaven and earth. This makes even pre-human natural processes essentially 'historical': not mere events, but more than that, creative acts.

One might object that God's thoughts and actions do not resemble that of finite man, and Miller would certainly answer that *his* religion is not without revelation. He would probably have agreed with Gillispie's verdict on his view: 'this may not be science, but it is Christianity'. Methodologically this self-taught quondam stonemason made no separation between them. Indeed, he published his ideas first of all in *The Witness*, a weekly of the Free Church of Scotland, in which the reader got theological and scientific instruction at the same time.

However much aesthetic and religious aspects may be recognized in nature, and however much they may influence scientific beliefs, scientific methodology nowadays does not admit them: even Pascal and Boyle, both fervent apologists of the Christian religion, nevertheless kept them out of their purely scientific works. **Question 4**

2.4 Compromises in system and method

Although uniformitarianism is now the scientific orthodoxy to which the first page of many a geological textbook pays due homage, it has been obliged to admit a number of qualifications under the coercion of facts. When it is said that operations in the past ensued from causes working with the same intensity as those active in the present, the meaning of 'the present' is ambiguous. It cannot stand for this very day (there might be no volcanic eruptions anywhere today), but it could stretch at least as far as human tradition reaches. The German geologist K. E. A. von Hoff (1822; 1824; 1834), who demonstrated that, as far as human recollection goes, all geological changes were of the same kind and degree as those going on now, was highly praised by Cuvier (who is generally considered as having used a non-actualistic method and as having arrived at a non-uniformitarian system). At any rate even if 'the events passing within a certain span of time' are repeated in a following period, there may be within each period (regarded as 'the present') a certain diversity of conditions, a 'history', just as there was 'history' (a certain series of different subsequent events) within a world period of ancient Stoic philosophy.

Even catastrophes *within* such a period would then not seriously damage the uniformitarian picture, as Whewell (1831) already maintained against Lyell. In practice this would efface the distinction between uniformitarians and catastrophists, at least in principle. In this way modern geologists who profess a uniformitarian faith have nevertheless been able to put forward a cyclical conception of earth history, the so-called 'pulse of the earth' (H. Umbgrove, H. Stille, L. Kober). For each period their systems are 'historical', whereas the overall scene is rather a-historical. They admit periods of relative inactivity in mountain formation, which alternate with periods of high activity. Though there may be similarity in 'kind' of geological operations, there is no uniformity in degree within each cycle.

When these cycles, though showing resemblances, are not wholly similar in character, the sequence of events bears an even stronger 'historical' character (the image used for it is that of motion in an ascending spiral). As C. C. Beringer (1951) put it: 'these epochs, in the inorganic as well as in the organic kingdom, are thoroughly unique (*einmalig:* they happen only once) and individual'.

This conception bears some resemblance to the theory of Giambattista Vico (1725), who held that certain patterns of development occur within different civilizations, but who superposed on these repeatable cycles of perpetual 'ideal history', the diversifying influences caused by human arbitrariness.

Another circumstance weakening the uniformitarian picture is that it has turned out to be impossible strictly to apply the principle of uniformity to the earliest geological epochs. When, supposedly, there was a primeval oxygen-free atmosphere, the processes of life and of the erosion of rocks must have been widely different from those actually in operation in an oxygenous atmosphere. There must have been then a 'pre-actualistic' period (Rutten, 1962).

Even to later periods, however, the actualistic method must be applied with reservations. In the first place it is generally recognized that plants and animals appeared on the scene rather late, and the geological surface phenomena must have been changed by their influence. Desert formation after the rise of vegetation must have been quite different from that before (J. Walther, 1893). In the second place, even beneath the surface of the earth not every kind of rock could have been formed at all epochs (E. Kaiser, 1931).

Here again a comparison with human history presents itself: unprecedented situations arise in the course of time, and these in their turn influence the further course of events, so that a mere repetition becomes impossible. **Question 5**

2.5 Evolutionism

Evolutionism, too, can be considered as a kind of compromise. Lamarck (1802; 1809) and Darwin (1859) energetically rejected catastrophes in geology and biology. As their belief in transmutation of species could hardly be supported by the direct evidence of actual events, they had to resort to other arguments. Darwin referred to the palaeontological record, which his opponents, the progressionists, regarded as the main support for *their* theory. To account for the fact that we never witness a transmutation of species in the present, the proponents of species transformation had to assume that it was an extremely slow process.

To the ideas of slowness and continuity of change Lyell's uniformitarianism gave powerful support. Though Lyell had stated that the progressionists, with their idea of historical development from lower to higher forms, were close to the transmutationists (Lamarckists), the 'transmutationist' Darwin himself felt more affinity with Lyell than with the Buckland school, because the latter could not provide him with the long time and the continuity he needed for his own system. Nevertheless, evolutionism —the idea that the animal kingdom slowly gave rise to higher forms through 'descent with modification' from lower forms in a continuous, non-cyclical sequence—bore (like progressionism and unlike uniformitarianism) a distinctly *historical* character. In geology the early evolutionists remained orthodox uniformitarians in the Lyellian sense; on the other hand, Lyell was (though not without some misgivings) converted to the Darwinian standpoint as to the organic world. Admitting evolution through descent with modification, he abandoned his steady-state theory for the organic world. Consequently, in order to explain the ascending series of animal forms, the evolutionists did not revert to *geological* causes: the historical element of their theory could not be supported by an a-historic geological system.

According to Darwin variation and natural selection give rise to new animal varieties, which, by multiplication of small effects throughout the ages, become different species.

The question then arises: how could this historical, irreversible and unique sequence of evolutionary stages be interpreted within the frame of uniformitarianism, which is typically a-historic? This could be done by a subtle change in the meaning of the expression 'what actually happens'. Large change in the past results from an accumulation of exceedingly small, undirected 'variations' (changes randomly going towards either 'higher' or 'lower' forms). Nevertheless it was held that the sum of these small changes altered the whole animal kingdom in a certain direction, in general from 'lower' to 'higher' forms. Consequently, it is not the average state of the animal kingdom itself, but the *change* in this state which remains constant. The changes go on slowly and continuously, at a uniform rate, by the same kind of causes (variations), but their effect is a more or less uniform *ascent*. There is uniformity of *coming-to-be* instead of uniformity of *being;* uniformity of the change of state, instead of uniformity of the state; or perhaps one could better say: *change* is now a property of the *situation*.

Neither Cuvier nor the progressionists had a scientific explanation for the new animal forms that appeared on the scene after each of the supposed catastrophes. The French zoologist Etienne Geoffroy Saint-Hilaire (1825; 1833), however, attributed biological transformations to change in the geological and climatological environment. His geological catastrophism was connected with a biological catastrophism. Each great change in the geological and climatological conditions causes a saltation (a large change) in the animal type, whereas during periods of tranquil geological change, the change in the animals, too, was small and gradual.

It is interesting to note that zoologists and geologists who believed in sudden catastrophic changes had in general some confidence in the corroboration of their ideas by means of laboratory experiments serving as models. Such experiments take, geologically speaking, an exceedingly short time and thus are highly 'catastrophic'. The more uniformitarian geologists and zoologists, however (Hutton, Darwin), had no great expectations from such short-timed experiments.

Question 6A
Question 6B

3 Historical and Physical Causality in Nature and History

The a-historic uniformitarian way of thinking has a tendency to shape each department of learning after the model of physics, reducing all change to the result of unalterable causes. In geology however, it had to give room to more historical conceptions.

Much confusion is caused by the fact that in English the term 'uniformitarianism' is used as a translation of the continental term 'actualism'. We have introduced (1958; 1963) the distinction between 'uniformitarianism' (a system) and 'actualism' (a method). The actualistic method tries, as much as possible, to explain past events by causes now in operation, but it does not dogmatically assert that this will always be possible, or that it is possible in every respect. It is very elastic: if applied in the most rigorous interpretation, it must lead to a uniformitarian system in the sense of Toulmin (see Unit 11, p. 65), but otherwise it may go together with the compromises expounded above. It is even an *indispensable* method: how could one reconstruct the past if there were not the slightest analogy to the present? Geology, and 'historical' natural sciences in general, would become a collection of fairy tales, full of the 'marvellous'. We cannot try to reconstruct the past without the help of unalterable physical and chemical laws, and we have as far as possible to assume an analogy to geological phenomena occurring today. The supposition that the physical laws (laws of gravitation, collision, etc.) have always been the same (that the elementary processes that build up the visible phenomena do not change), is at the basis of *all* geological systems. In this sense, then, both uniformitarians and non-uniformitarians apply an actualistic method. The non-uniformitarian, however, maintains that not all coincidences of conditions must be repeatable (or that at least they may be highly improbable), so that events like those ensuing from a certain coincidence probably will not happen a second time.

However, the principle of actualism holds not only that elementary physical and chemical processes are always the same, but also that some of the complicated combinations which determine the general *geological* phenomena (erosion, sedimentation, etc.) have also been always the same. It 'de-historicizes' these combinations by presenting them as repeatable and in fact often repeated.

The non-uniformitarian will proclaim his adherence to this actualistic method; he will maintain the identity of physical causes in past and present, and also, as far as possible, that of the more complicated coincidences of physical and biological conditions; but he will if necessary recognize that not every rock could take rise at every time, and that, when certain types of animals appear for the first time, this is due to a once-and-for-all meeting together of certain causal series, and in this respect bears the typically 'historical' character of uniqueness.

The English philosopher and historian of science William Whewell (1840) made a distinction between 'historical' causality and 'mechanical' causality. 'Force', he said, 'is the cause of motion . . . at all times and under all circumstances . . . but the cause of the present condition and elevation of the Alps . . . was manifested in a series of events of which each happened but once, and occupied its proper place in the series of time.'

In fact, every event in the history of nature has its physical as well as its historical aspects. The former relate to its actualistic, repeatable elements, the latter to the unrepeatable, unique whole. The former are determined by 'physical' causality, the latter by 'historical' causality. The more complicated the phenomena, the more individual and historical they are, the less likely they are to occur again.

Mineral genesis requires certain particles to arrange themselves in a definite pattern; rock formation requires several minerals to unite in a peculiar structure; in mountain building rocks partake in certain events so that special aggregates and shapes arise. The origin of living beings is so complicated that it has not been reproduced, and even if this were possible, there would be no guarantee that nature originally wrought them in the same way.

It is, however, with *homo sapiens*, with his intellect and will, that the historical character of events becomes most evident. The French philosopher A. Cournot (1872) considered 'historical chance' as the accidental coincidence in a certain phenomenon of several mutually independent series of causes. This fortuitous character does not imply that there are effects without cause: the series may be strictly determined, and even their meeting together may be determinate. Besides these 'fortuitous' events there are events that are recognizable as consequences of the immanent necessity of a sequence of causes and effects. For example, if the human race were to perish through collision with a comet, this would be an 'accidental' event; whereas the extinction of living beings as a result of gradual cooling down of the earth, would be a necessity.

In human history Cournot distinguished between the slow development of society in the course of ages by general immanent necessities, and local, fortuitous shocks, which do not essentially change the social situation. In spite of his predilection for phenomena of the first type, Cournot had an open eye for what Pascal meant by saying that 'if Cleopatra's nose had been shorter, the whole face of the earth would have changed'. He recognized that one 'fortuitous' event, in the history of the earth as well as in political history, could have influenced all later events. The resemblance between the development of nature and human history is, in his opinion, manifest in that both have arrived at a similar 'modern' phase. In geology the forces of construction and destruction have abated and the capacity for change of organic types has disappeared, so that we have reached the final stage of a series that was hitherto progressive; in future there will be only oscillations about a stable average state. In human history there is an analogous situation: since the fifteenth century a state of equilibrium has been establishing itself, so that the influence of politics (i.e. that of persons, 'history in the proper sense') decreases and that of science and industry (i.e. that of events determined by social and economic laws) increases, until its progression ends in a state of equilibrium. Thus in the history of nature, as well as in that of mankind, Cournot considers important what Whewell termed 'historical causality', but for the modern era he (like Cuvier before him) accepted in both fields a course of a more uniformitarian character, subject to definite 'scientific' laws.

Cournot's ideas greatly influenced J. B. Bury (1916), who said that a historical contingency is 'the valuable collision of two or more independent chains of causes'. As time goes on contingencies will become less important and chance will have less power over the course of events. Social development tends to become more logical, owing to an increase of man's command over his own nature and over his social and political situation, which will be moulded less and less by single individuals. Like Cournot, Bury believes that contingent passions and individual opinions will have less influence and that mass forces will prevail; statistics will state *laws* of a scientific character.[1]

We have considered the unrepeatable, unique event as typically historical and our conclusion is that nature has a 'history'. In the history of nature, as in that of mankind, the historical uniqueness results from the coincidence of diverse causal series. But whereas on the one hand nature possesses a historical aspect, on the other hand human history shows some 'scientific' aspects of uniformity and actuality. And in human history, as in that of nature, the repetition is in the elements rather than in the combinations. Each historical description uses notions like monarchy, democracy, conspiracy, revolution, dictatorship, and although these may not have wholly identical contents in different epochs, there is enough analogy to justify the term 'dictator' for Sulla as well as for Stalin. Events, situations and human feelings of the present are used for describing the past. The use of the same terms implies that there is an affinity great enough for us to understand the past by means of analogous phenomena and motives taken from the present. One would not be able to understand the as yet unknown without using analogies borrowed from the known. Though in a certain sense history *does not* repeat itself, yet there is sufficient resemblance between component phenomena of the present and the past to conclude that history *does* repeat itself. There are no longer conspiracies against a dictator of the Roman republic, but there are still conspiracies against dictators.

This repetitiveness is perhaps even more true of the inner side, the thoughts motivating the people who play a role in historical events. These, too, are repeatable components of history, as physical forces and processes are repeatable components of geological phenomena. The character of these components, however, makes evident the difference between the historiography of mankind and that of nature. Historiography of man must be more than a registration of events: it must describe *actions*. The historian has to re-live what the people of the past felt, thought and did and he has to analyse this re-enactment. This pretention can only be justified by the assumption that the fundamental motives of human thought, will and action (love, hatred, ambition, avarice) observed in ourselves and in our contemporaries also moved and motivated our ancestors. When a certain situation stimulated certain feelings, thoughts and actions in persons of the past, we assume we can re-live that situation and the spiritual actions and reactions it aroused because our mind can react in the same way, if not in fact, then in sympathetic imagination, just as it does in understanding our contemporaries. On this actualistic pre-conception all historical interpretation is based.

When it is said that history does not repeat itself and that history does repeat itself, these verdicts do not have the same meaning for the history of mankind and the history of nature. It is only a formal resemblance and not a true similarity. There is an analogy in the pattern (the unique wholes and the repeatable components), but the sentiments of hatred and love of diverse people from diverse epochs (and even the revolutions and other historical events of different epochs) cannot be considered identical in the same way as the collisions of elastic bodies in the past and the present. The historiography of mankind deals not only with events which people *undergo;* it deals in particular with acts they *commit*. Besides, whereas palaeontology describes a history of the species, human history is also one of *persons* who, either independently or as part of a collectivity (nation, party, class), consciously contribute to the steering of the course of events in a particular direction. Only man 'makes' history, and this makes *his* history qualitatively different from the history of the earth and the animal world: a new generation is not a continuation of the older one in the same way as it is in sub-human nature.

Question 7

[1] It is strange that Bury did not acknowledge what he owed to Cournot, as R. G. Collingwood remarked (*The Idea of History*, 1946). But in our opinion, Collingwood was wrong in thinking that Bury had misunderstood Cournot.

4 The Integration of History into Nature

4.1 Darwinism and history

The principal mechanism of evolution, Natural Selection, is in Darwin's view almost independent of geological and climatological conditions, but it is determined by the biological milieu. In this way tension is avoided between his uniformitarian geology (which does not imply a 'development') and his essentially different quasi-uniformitarian biology (which does admit development).

Change within the organic world takes place at a more or less uniform rate: 'The amount of organic change in the fossils of consecutive formations probably serves as a fair measure of the lapse of actual time'. Yet 'we must not overrate the accuracy of organic change as a measure of time': migration may modify the rate of change, which according to Darwin was probably slower during early periods of the earth's history. (This opinion offers an interesting contrast to theories proclaiming that in the earliest periods the world was more pliable, and geological and biological changes more rapid than in more recent times.)

Like the progressionists before him, Darwin introduced a historical aspect into biology (' . . . when we regard every production of nature as one which has a history . . . our classifications will come to be . . . genealogies'). But by considering organic change as uniform, by submitting it to 'laws' and explaining it mechanistically rather than vitalistically, he had given it a 'scientific', physical character: a-historic physics was to Darwinists the ideal true science.

In contrast to Charles Lyell and to A. R. Wallace (who put forward the theory of natural selection independently of Darwin), Darwin considered the origin of human reason and culture not as the birth of a new order, but rather as a consequence of the same 'law' of uniform development that was behind all previous progress. He made it an essential point that no discontinuity exists. Man's psyche, his reason and his social instincts have been gradually developed by natural selection.

Darwin's disciples in prehistoric archaeology, too, considered human history as a prolongation of the history of the animal world; and they extrapolated the idea of evolution from the area of nature to that of human cultural progress. The idea of progress in history, however, had been applied in human historiography long before the evolutionary theory was accepted in biology.

In Darwin's conception the term 'evolution' is not very well chosen. Darwin emphatically rejected the Lamarckian idea of a 'tendency of nature' towards higher forms through the 'power of life': 'the tacit assumption . . . that there is some innate tendency towards continued development in mind and body . . .'. Natural Selection 'acts only tentatively'; it does not imply an upward movement, but only a transformation, either for better or for worse, improvement *or* degeneration.

Despite this, Darwinism is generally considered a theory of progress, and it is quite easy to find support for this interpretation in Darwin's own writings. In particular the branch of primates ending in *homo sapiens* is considered by him as a progressive series, in which mental development is still undergoing a further evolution.

> Man may be excused for feeling some pride at having risen, though not through his own exertions, to the very summit of the organic scale, and the fact of his having thus risen, instead of having been originally placed there, may give him hope for a still higher destiny in the distant future (C. Darwin, *The Descent of Man*, 1871).

In this principal series of development the history of nature merges without any discontinuity into the history of mankind: from amoeba to primitive man, from primitive man to civilized man and finally to superman. The history of man shows in this view the same features as that of the rest of the animal world; in both there is progression and ramification. The primitive peoples still existing have been left behind; either they have not yet gone through the development within the main line, or they are on a sideline. Even now there exist some races that are inferior, and therefore doomed to destruction; but the main line ascends towards the civilized nations ('The western nations of Europe, who now so immeasurably surpass their former savage progenitors, and stand at the summit of civilization. . .').

Now Darwin recognized that direct proof of the transformation of animal forms does not go further than 'varieties', which he interpreted as 'incipient species'. Though, in principle, natural selection effects progression as well as retrogression (it is an undirected cause), progression prevails: 'Progress has been much more general than retrogression'. But if progression does not necessarily follow from variation and selection, how does he know that it prevails? The palaeontological record gave support to it, but the many gaps in the series made the proof insecure. Embryology therefore was 'the weightiest argument of all' to Darwin. According to the thesis that the subsequent stages of development of the phylum must be repeated in the development of the embryo, the palaeontological sequence must present a genealogical series. ('The embryo in the course of development generally rises in organization . . . no-one probably will dispute that the butterfly is higher than the caterpillar.') The reserve Darwin now and again shows with regard to the progressive direction of natural selection ('natural selection acts only tentatively') is more than compensated by verdicts to the contrary: 'Natural selection works solely by and for the good of each being, all corporeal and mental endowments will tend to progress towards perfection'; 'variation will cause the slight alterations, generation will multiply them almost infinitely, and natural selection will pick out with unerring skill each improvement'.

Though Darwin propounded his historical conception on grounds borrowed from palaeontology and embryology, the 'explanation' of the 'origin of species' had to be given by a-historical natural selection: 'a-historical' because in the struggle for life among the countless undirected variations those individuals are preserved which can best maintain themselves in a certain milieu; and this again does not imply 'direction'. Consequently Darwin had to inject progress into the non-directional natural selection, and it is often said that there is 'advance', 'progress', 'improvement' and 'perfection' by natural selection. In spite of this agent's being a blind force and in spite of the rejection of teleological arguments, value judgements (which are a characteristic of *human* historiography) emerge again and again in Darwin's interpretations.

Through Darwin's theory of progressive evolution a historical element entered in a modest way into the science of nature, but at the same time a 'scientific' element strongly penetrated into human historiography. Biology first borrowed a certain view from history, and then this historicized science of nature proclaimed history itself as a continuation of biological evolution. It seemed as if the historical view had gained quite a success when nature was considered as subject to a historical process. But the procedure was reversed, and human history was regarded as subject to the same natural laws as the historical process of nature herself. The methods and theories of evolutionary biology were declared applicable to the historiography of mankind.

Biological evolution, however, takes many thousands of years to effect a change of some importance, whereas significant developments in human culture and society cover at most a few thousands of years, and often only a couple of centuries or decades or years. **Question 8**

Moreover, the objects of biological evolution are *subjected* to 'development', whereas human beings to a large extent *co-determine* the course of development. Men have a memory of past generations and they have distinct ideas and intentions about future events. These memories and plans largely determine their actions. Final causes, contemptuously discarded from biology by many of its cultivators, play a predominant role in human life. In our view of non-human nature no ethical and aesthetic values are acknowledged; man may attribute a 'good' smell to some chemical compounds and a 'bad' smell to other ones; he may consider some flowers beautiful and others less so, and he may dub some animals 'pests' and others beneficial; but these characteristics have no legitimate place in the scientific world picture. The only 'final cause' that counts in biology is that of the animal's maintaining itself in the struggle for life. To use the term 'progress' for biological evolution is legitimate only if it means that one specific form ensues from a different previous one; but it should not tacitly include (as it usually does) the idea of 'improvement', since this implies a standard of values (*viz.* the idea that evolution goes upward towards Man who as Darwin says can be proud to have *risen*, though not by his own efforts, from a low state to the highest form of life).

However, 'progress' and even 'improvement' are often considered laws of nature, so that the 'tendency of Nature' to improvement creeps in. Moreover as historical progress is held to be a mere continuation of natural progress, human history (in particular social history, which is less dependent on individuals) is considered a necessary outcome of a natural law. Consequently, not only is Man the highest product of Natural Selection, but this ascent is often believed to be still going on. Each earlier stage is valued mainly as preparatory for the next stage, which possesses a higher value: e.g. the bourgeoisie may be appreciated only in so far as it prepared the way for socialism. Following the same trend of thought, Darwin agreed that the whole of history served only one *purpose*, namely that of finally producing the best of all cultures and nations:

> Looking to the distant future, I do not think that the Rev. Mr. Zincke takes an exaggerated view when he says: 'All other series of events—as that which resulted in the culture of mind in Greece, and that which resulted in the empire of Rome—only appear to have purpose and value when viewed in connection with, or rather, as subsidiary to . . . the great stream of Anglo-Saxon emigration to the west.' Obscure as is the problem of the advance of civilization, we can at least see that a nation which produced during a lengthened period the greatest number of highly intellectual, energetic, brave, patriotic, and benevolent men, would generally prevail over less favoured nations (Darwin, 1871).

The notion of 'precursors' is clearly stated in the above quotation: they have only 'purpose and value' in so far as they are necessary to the rise of *our* modern perfection in political, social, scientific, or philosophical respects.

Now if there is any realm of culture in which one may justifiably speak of historical 'progress' and 'improvement' it is that of natural science. Small wonder, then, that the notion of 'precursors' flourishes in *its* historiography. Precursors are people whose only significance seems to be that their ideas and discoveries come back in a more elaborate and correct form with later scientists of greater fame. A precursor, a herald announces the king's arrival and he exists (as a herald) not for his own sake. In our view, this tendency is deplorable. Each scholar, however, even if he may appear to be insignificant in our eyes, should be considered in his own right and not as existing for the sake of those who live after him. This implies that he should be seen as much as possible in the light of the factual knowledge, the beliefs, the means of research and the social situation of his own time and place. Only after that should we view him in retrospect, asking whether, how and why he paved the way for the present situation.

Though there surely is progress, in the sense of 'improvement', in science, it seems problematical whether this has anything to do with an 'evolution' of man. The process takes place consciously and deliberately (however much the outcome may often have differed from the intentions of the scientist). Science bears a cumulative character: the experience and knowledge of former generations are delivered to the next one, so that even when this may be intellectually inferior, it still possesses a greater knowledge of nature. Einstein knew more of physics than Newton, but we do not for this reason consider him a greater physicist. There is progress and improvement in mathematics, but nobody says that Gauss is a greater mathematician than Archimedes. As Newton reminds us, we see farther than our forebears, not because we are greater, but because we are standing upon their shoulders.

As to ethics, there is reasonable doubt whether we are on a higher level than primitive tribes; the level of ethical attainment is perhaps connected with individuals rather than with nations.

'Aesthetically we may admit no progress.' 'No agreed standards exist for judging the quality of thought' (Childe). If there be any improvement in the quality of art, one wonders why many modern people admire 'primitive' art and even imitate it after we have gone through the refinement of the Renaissance. The Hymn of the Sun of Akhnaton cannot be said to represent a lower stage of poetry than a poem of Keats.

We do not add to the ethical insight or the aesthetic creativity of a former generation as we add to its technological and scientific knowledge.

The term 'evolution', then, should be avoided outside the realm of biology. Neither chairs and carriages, nor parliaments, nor art and religion (and not even science) 'evolve' by a 'natural' process, and within a few years. There *may* be improvement and there *may* be novelties in particular fields (scientific theory; knowledge of society; technology); but it is just a non-scientific *dogma*, a 'religion without revelation', that the historical process is 'a mere extension' of natural processes. The conception of a law of progress and continual development is thus 'a mere confusion of thought, bred of an unnatural union between man's belief in his own superiority to nature, and his belief that he is nothing more than a part of nature. If either belief is true, the other is false' (Collingwood, 1936).

A strong reaction against the 'physicalization' of the history of nature and the history of mankind flourished among German geologists, palaeontologists and archaeologists of the second quarter of the twentieth century, who tried, the other way round, essentially to 'historicize' nature. To them it was not enough to underline the unique character of the history of nature; they emphasized that this history was 'really historical' in that it bore the same character as the history of mankind in its course, its structure and its meaning (K. Beurlen, 1936; E. Becksmann, 1940). The individual character of each geological constellation was now even termed a '*personal* character' (E. Wahle, 1940–1); the history of mankind was said to be just an episode in 'the unique immense history of the world' (C. C. Beringer, 1954).

The last sentence demonstrates that this anti-Darwinistic and vitalistic view, through its annexation of nature by history, led to the same result as the annexation of history by nature, both ending in all-embracing naturalistic systems and both being based on naturalistic beliefs. **Question 9**

4.2 Prehistory

The field of prehistoric archaeology borders on one side that of the science of palaeontology and on the other side that of the 'ordinary' historiography of mankind. The mixing up of scientific and historical thinking will most easily occur in such an area, where neither the scientific nor the current historiographical methods can operate satisfactorily.

Originally this field was confounded with that of palaeontology. The prehistoric archaeologist finds remains of men and their culture (dwelling sites, burial places, instruments, ornaments) and tries to deduce from these relics the migrations, ways of life, means of subsistence and religious rites of these people. He uses the fundamental rule of stratigraphical geology: the upper site is in general considered to be the most recent one, and the 'leading fossils' of geology find their counterparts in the 'type-fossils' of archaeology. In geology certain fossil genera or species are considered as characteristic for a certain epoch; in archaeology certain relics of culture (e.g. axes) are considered as characteristic for the stage of development of the community that produced them. Neither the palaeontologist nor the prehistoric archaeologist encounters the written word, the direct symbol of thoughts. Instead from the remnants of things both arrange their finds in chronological series.

Just as the evolution from lower to higher animal types was considered to follow a chronological series with increasing differentiation and efficiency, it seemed plausible to suppose a progression in the archaeological finds and to assume that particular implements, such as axes, 'develop' towards greater differentiation and efficiency. The more perfect forms were held to expel by a kind of natural selection the less adapted ones; consequently, they must be of a later date. Thus what determines the chronological series is what the *modern* scholar deems greater efficiency. The actualistic principle here is that efficiency was the only purpose of the makers, and it is forgotten that this same evolutionary theory holds that the prehistoric users may not have had the same ideas of efficiency as nineteenth-century technicians: e.g. they may have attributed great magical efficiency to a certain shape of their tool.

Moreover, one should realize that *palaeontological* relics of developing animal species are themselves a goal of research, whereas *archaeological* relics are but *products* of developing groups of people. Consequently it is not those objects themselves but the thoughts and acts of their makers and users which are the final aim of the investigation (V. G. Childe, 1956). The investigator of prehistory, who has no personal testimonies of human beings but only artefacts to deal with, is in danger of forgetting the makers because of his concentration on these artefacts. By his method of investigation and by the character of his data, he is even more liable to follow a naturalistic way of thinking than the student of political history. He has absorbed the Darwinian ideas of continuous development and he constructs typological series of the development of implements in the same way as the palaeontologists arrange fossils in an evolutionary series from eohippus to horse. As a Scandinavian archaeologist has put it:

> Typology is the application of Darwinism to the products of human labour. It starts from the assumption that human will is subject to certain laws, analogous to those valid in the development of the organic kingdom. The antiquities develop themselves as if they were living organisms (N. Aberg, 1929).

It is true that the 'living will of the maker' is taken into account; but as this will is subject to 'development according to law', even this historical element acquires the character of a purely scientific law.

We have no right however to criticize a verdict that 'antiquities evolve as if they were living organisms', as long as we ourselves speak about 'the development of the chair' or 'the evolution of the vehicle'; as if the coach existed to bring forth the brougham and the brougham begot the automobile. This establishes the wrong idea that there is some immanent urge to development, as if one form *must* necessarily give rise to the next one, as the caterpillar must turn into a butterfly.

In the cultural history of the recent period such a separation of cultural objects from their makers and users is even less excusable than in prehistoric archaeology. If one has no other data than products of technical skill, it becomes indeed very difficult to give a reliable picture of the whole area of thoughts and acts of a past generation. He who wants to understand the real thought of a neolithic man must try to become one himself in his mind; and if he cannot or does not wish to do so, he must be contented with arranging in series the skulls, implements and drawings of this people (B. Croce, 1915).

Those Darwinistic archaeologists who felt the lacunae in this *Gegenstandskunde* (science of objects) reacted by reverting to ethnography. The first plate in Lyell's *The Antiquity of Man* (1863) shows a village built on piles in a Swiss lake, after an 'ideal restoration', partly based on 'similar habitations' of Papoos in contemporary New Guinea.

The supposedly logical series of stages of development of religion (from animism to monotheism) and of sexual behaviour (from promiscuity to monogamy) were transformed into chronological series. In this way the scanty data have been filled out into a vivid picture embracing the fullness of past human thought and action. Nevertheless, such theories too bear a strongly naturalistic character, since 'cultural evolution' is considered just an extension of biological evolution. Cultural forms, however, may have been instituted and maintained for reasons that cause deviation from what *we* consider a 'logical' and 'natural' development.

Prehistoric archaeology works in unfavourable conditions as compared with ordinary historiography. In the latter the historian is better able to penetrate into the inner core of the events, into the thoughts and feelings of the actors. There are written documents containing information on thoughts and acts, so that history in the full sense may be reproduced. In prehistoric archaeology such a 'history' does exist (at least if we assume that actors belonged to the species *homo sapiens*); but data are too sparse to allow it in the full sense to be written. The archaeologist is hardly confronted with individuals. Each new type of implement may have been the result of a unique creative act of a particularly intelligent individual, but only those inventions have been preserved as archaeological data which have been accepted

Figure 3 A village built on piles in a Swiss lake, from Charles Lyell, The Geological Evidences of the Antiquity of Man, *1863.*

and imitated by some community (Childe, 1956). As the palaeontologist recognizes in the relics of an individual animal merely those of a representative of a whole species, the archaeologist recognizes in a cultural relic a specimen of a multitude of similar objects, each expressing the activity of an individual who represents a whole community of similar individuals. Technological activities of a group can thus be reenacted to a certain extent, but for the more 'spiritual' activities any reconstruction will be uncertain. Here *contemporary* tribes are supposed to represent earlier stages of development through which civilized nations have already passed.

This archaeological–ethnological actualism was advocated by the famous American anthropologist Lewis H. Morgan in his *Ancient Society, or Researches in the Lines of Progress from Savagery through Barbarism to Civilization* (1877).[1] Development through these stages is said to have been running 'in different but uniform channels upon all continents, and very similar in all tribes and nations of mankind down to the same status of development'. The rate of development may be different under different circumstances, which explains why (though mankind is considered one in origin) earlier stages still exist here and there. In this way Europeans find their remote ancestors represented in Australian aborigines in the state of savagery, and in American Indians in the state of barbarism. Ethnological material is considered to be also archaeological material; the chronological natural sequence is considered also the logical one. Morgan builds up his archaeology without needing direct archaeological data, just by projecting ethnological data back into the past.

The causal relation between subsequent technical skills Morgan terms 'progressive' (they are external; the earlier gives rise to a later one), that of social institutions he calls 'unfolding' (they are immanent; from 'primary germs of thought' present in the state of savagery, institutions were developed in the state of barbarism; 'they have a lineal descent through the ages, with the streams of blood, as well as a logical development'). Whence does this uniformity of cultural development arise? From the similarity of the human brain in all races; the whole development was 'predetermined' from the start, which was identical for all races.

> The evolution of these germs of thought has been guided by a natural logic which formed an essential attribute of the brain itself. So unerringly has this principle performed its functions in all conditions of experience, and in all periods of time, that its results are uniform . . . . But from a few germs of thought, conceived in the early ages, have been evolved all the principal institutions of mankind.

[1] This work greatly influenced Friedrich Engels' *Der Ursprung der Familie, des Privateigentums und des Staats in Anschlusz an Lewis H. Morgan's Forschungen*, 1884.

The external conditions hardly caused any divergence of development, except with respect to the tempo of change. It is this that leads Morgan to see existing primitive tribes as 'living fossils' of our own ancient institutions and customs.

Evidently Morgan's 'natural logic of development' purports to be a natural law of development. There is a close resemblance to Darwin's conception of embryological development. It is a *history* of mankind, but this implies only that the development goes through a continuous, irreversible evolutionary series of stages, and it is also a 'natural' development, almost identical for different tribes.

The core of Morgan's theory is the principle of the uniformity of development of the human mind. His method is undoubtedly actualistic in that he borrows the picture of former stages of development of our Western civilization from still existing 'backward' societies. In his conception even a real historical understanding of them is possible, because our minds, though further developed, share a common origin with theirs: 'A common principle of intelligence meets us in the savage, in the barbarian, and in civilized man.'

Though Morgan does not explicitly refer to biological parallels, some of his conceptions closely resemble those of Lamarck. His 'natural logic' that necessarily urges the development from germs already present in savages, reminds us of Lamarck's 'tendency of nature' which causes a linear ascent in the animal kingdom. Morgan bases his chronological series on human communities still existing, without using archaeological finds; Lamarck's series of development is based on a 'natural' sequence of still existing animals, without making much use of palaeontology. In both cases 'history' is obtained by proclaiming a 'logical' sequence of now living species to be also a chronological sequence of descent, a method typical of a science of nature but not typical of historiography proper, which considers the relics of the past as its foundation.

Archaeologists of note have criticized the general laws of natural development (from hunters to nomads to cattle-breeding and agricultural societies), pointing out that each natural environment co-determines the way of life. They have criticized also the typological series, which assume laws of development that do not take into account the changing conditions of life, e.g. because of changes of climate ('The proud chronological construction on the fundament of typology is a dangerous illusion', A. W. Brøgger, 1926).

The well-known archaeologist V. G. Childe (1956), though accepting that contemporary primitive African and Australian tribes offer 'accurate parallels' to the older documents of European prehistory, rejected Morgan's idea that the development of culture has advanced everywhere on parallel lines. In his opinion even the identity of the course of technological progress is not confirmed by archaeological data ('Nor is it really practicable to establish an identity between the technology of any contemporary tribe with that of any prehistoric community'). Typological series he deems unreliable ('the neatest typological series may turn out to be bogus'). While recognizing that on the whole successive cultures show technical progress, he criticizes the 'natural historical, darwinistic-spencerian' methods in archaeology of those who 'looked upon the development of human culture as a continuation of organic evolution' and confined human history 'within the categories of natural history', inventing series illustrating 'an "evolution" from clubs to howitzers . . . from magic to monotheism, from waist strings to trousers'.

In Childe's opinion the archaeologist is a historian and he has to re-enact in his own mind the thought of prehistoric man, as 'realized by a society of persons'. He recognized, however, that here we meet with a great difficulty: the way in which reason works is not immutable; it has varied with time and place. Yet to those who would say that prehistoric men had a different conception of the world than we, and that this cannot be recaptured by considering implements, ornaments, etc., Childe answered that this negative conclusion reflects a misconception of the function of rational thought and of the nature of scientific laws, as 'the function of reason is not contemplation, but action'.

He deemed it a fortunate circumstance that archaeological finds testify only to the technological actions of their makers, and that only the thoughts they had in their mind when making them can be re-enacted. It is recognized that, when making them, they had more in their mind, *viz.* rites, conjurations and sacrifices. But thanks to technological and scientific progress we *now* know that these are quite irrelevant to the success of the operation, being but 'futile accessories, expressive of ideological delusions'. Though these ideological errors and superstitions gave rise to the building of temples and tombs, which are important archaeological data, it is in his view no serious lack that we can re-enact only the technological parts (the working precepts) of prehistoric thought; for 'their real historical function need have nothing to do with the subjective intentions of the actors'. It consisted in the practical science that was applied in building, in the economic role, in the distribution of wealth and in expressing social solidarity.

Childe, however, conceded that probably none of these (modern) aspects was present in the 'false consciousness' of the builders; they thought that they built the temples for their gods, but 'objectively' it was to reach an economic and social goal. 'Their "motives", like their emotions, have been lost for ever, just because they were illusions. Does that matter?'

What others deplore, namely that re-enactment hardly transcends the technological level, this archaeologist did not consider a loss, because most of this primitive thought is now considered 'objectively' irrelevant and obsolete. This is a strange kind of re-enactment: the goal of re-living the *whole* life and thought of prehistoric people is abandoned. Though recognizing that the 'categories of logic', the 'instruments' the archaeologist uses when observing ancient cultures, are not immutable, but that they themselves are part of his own, twentieth-century, culture, he boldly sorts out the 'subjective' intentions of the ancestors because (in his own subjective opinion) they were not objectively relevant and had no 'real historical function'. The empty space left by the lost motives (the 'false consciousness') of prehistoric people is filled up with the social and economic motives excogitated by the twentieth-century archaeologist, even though avowedly they were not present in the ancient people's minds.

In order to attain this same end, the restoration of archaeology as a part of 'history', some German archaeologists (e.g. E. Wahle, 1940–1) fell into the opposite error. Wahle opposed the naturalistic conception of a strict law governing development. He condemned the transference of biological notions (mutation, atavism, etc.) to prehistory as 'rationalistic-unhistorical thought'. The essence of historical, and consequently also of prehistoric, development is 'the outstanding personality', and we should look for traces of the 'authoritarian individual'. (This was written in Germany in 1941.) He recognized, however, that these personalities can only be guessed and not demonstrated: in fact prehistory is more familiar with culture provinces than with individuals. Great historic change took place at a rapid pace, so that the forces that made history must be sought in the (short) periods of transition which left behind almost no relics.

This is no great help: archaeology, as Wahle himself recognized, is forced to rely on 'knowledge of objects' (*Gegenstandskunde*), and yet it has to be 'based on periods virtually without objects'. He is realistic enough to see the difficulty; but, so he says, at any rate this conception demonstrates the limits of what can be reached, and it will further a more vivid image of the prehistorical past.

Wahle's critique of those who link prehistory too closely to a Darwinistic theory of descent and to natural selection is marred by the fact that he himself links it to a non-Darwinian theory of descent. He draws a parallel with the palaeontology of the German school of Beurlen, which also admitted long periods of quiet development alternating with short periods of violent explosive development in which a new 'plan of formation' arose. He considers geology, too, like the history of mankind, 'genuinely historical'; it bears a 'personal' and 'unique' (*einmalig*) character. Natural selection and 'adaptation' are in Wahle's opinion not sufficient for reviving the past; 'vital force' has to be introduced, and this must replace the mere interpretation of archaeological findings.

However, neither the Darwinists' mechanistic conception of history, nor the vitalistic reaction to it of this 'historical' school, makes a fundamental distinction between the phenomena of nature and the events of human history. With the latter school it is not physics, but biology, that provides the general model which the historiography of man also has to follow. The metaphysics is different: the rather clear notions of Darwinistic rationalism are replaced in that German school by an irrational mysticism of the 'form-giving force of life' (*gestaltende Lebenskraft*). But in both cases the distinction between the human and the sub-human has been made vague, and man is interpreted on the basis of the sub-human (the mechanical or the biological). This would-be 'historicism' is a naturalism in disguise; it is not conscious will and action that characterize 'history' in its full sense, but 'vital force', present in man, but also in animals, plants and even the earth. It is another vain effort to liberate prehistory from the scientistic delusion.

Our thought may follow lines differing from those of our forebears, but all historiography is based on the actualistic principle that their spiritual life to a large extent resembled ours, at least in its component 'elements', and that we are able to reconstruct their acts and thoughts with the help of these elements of thought and the relics of their culture. If these relics are too scarce, true historiography, and even chronicle-writing (the mere registration of facts), become impossible. Not in principle, but in fact, we are then incapable of wholly performing the task of historiography.

Historiography in the proper sense is impossible in those areas that are inaccessible to historical actualism, i.e. where not only the capacity but also the possibility of re-enactment is lacking because we are confronted with beings whose thoughts and feelings (if they had any) are barred to man. 'Do you wish to understand the true history of a blade of grass? Try to become a blade of grass; and if you cannot do it, satisfy yourself with analysing its parts, and even arranging them in a sort of ideal or fanciful history' (B. Croce). There is, however, a difference between the 'history' of a blade of grass and that of neolithic man. The latter was a 'historical reality'; if we can approach him only in a scientific way (by analysing or classifying him from an external point of view) and not in a historical way (by making his life our own), this is but a failure of our historical insight and historical data (Collingwood, 1946). **Question 10**

Conclusion

The *history* of mankind differs from the 'history' (quasi-history) of the solar system, the earth's crust and the animal kingdom in that it is not a mere sequence of events that can be registered but the sequence of actions of beings who are able to co-determine the events in which they play a role.

Consequently, the *historiography* of mankind differs from that of the earth (geology) or of living beings (palaeontology) in that the actors of the drama are of the same kind as the historiographers who reconstruct it. The historian not only describes the exterior of the actions of the actors, but (in principle) also understands their inner side. The finalistic question *why*, ejected from science as being too anthropomorphic, is paramount in human historiography. Sympathetic understanding, being in other people's skin, is possible only with regard to our fellow-men, and not with regard to stars, plants and animals.

Figure 4 G. W. Hegel, after Xeller (Mansell Collection).

Is there a *history* of the earth and of the organic kingdom? Yes, if we demand only that history be a unique, unrepeatable sequence of events. Hegel's verdict that in nature there is always repetition, whereas history shows progression, was perhaps acceptable before about 1800, but it is not so any more.

Is there a *historiography* of the earth and of the organic kingdom? Yes, if historiography means only the reconstruction of past events with the help of the relics of the past. Whereas the historiography of mankind uses documents, monuments and other relics of culture, the historiography of nature studies the earth's crust with its fossils. In both cases imagination controlled by reason has to re-enact the past events. But the sympathetic re-living, indispensable in human historiography (which is concerned not only with the outside, the mere events, but also with the inside, the motives of actions), is replaced in 'historical' natural sciences by a scientific reconstruction. Instead of psychology, ethnology, sociology, etc., it is then physics, chemistry, physiology that help to revive the past. Thus there is both a history and a historiography of nature 'left to herself', which by the unrepeatable character of many of the phenomena occurring in it, differs from the typical science of nature, *viz.* experimental physics. But this lacks an essential characteristic of what is usually understood by 'history'. History in the proper sense relates events connected with *homo sapiens*, the creator of culture. Historiography in the proper sense is possible only on the basis of data bearing witness to *or* betokening cultural activities. Human remains that cannot be connected with traces of culture are no historiographic sources. Even when originating from *homo sapiens*, they are of only palaeontological interest.

The historiography of mankind and that of nature have in common an actualistic trait. Thanks to a repetitive pattern of personal and collective thought in one case, and to a repetitive pattern of physical events in the other (that is, owing to the fact that 'history repeats itself') the proper task of the historian can be fulfilled, *viz.* the reconstruction of 'history that does not repeat itself'. **Question 11**

Author's Note

This unit is mainly based on the present author's *The Principle of Uniformity in Geology, Biology and Theology*, Leyden, 1958[1], 1963[2], chapters 1 and 2, and *Natuur en Geschiedenis*, Noord-Hollandsche Uitgeversmaatschappij (Mededelingen Kon. Ned. Akad. v. Wetenschappen, afd. Letterkunde, nieuwe reeks, deel 29, no. 9), Amsterdam, 1966.

Further Reading

A. Cournot, *Considérations sur la marche des idées et des événements dans les temps modernes*, Paris, 1872.

R. Hofstadter, *Social Darwinism in American Thought* (first edition 1944), revised edition, Beacon Press, Boston, 1955.

R. G. Collingwood, *The Idea of History* (first edition 1945), Oxford Paperbacks no. 27, Oxford University Press, 1961.

V. G. Childe, *Piecing Together the Past, the interpretation of archaeological data*, Routledge & Kegan Paul, 1956.

Part Two (by Clive Lawless)

Questions

The questions in this part of the unit relate to particular sections of Part One and their place is indicated in the margin. You should tackle the appropriate question at the end of a section before commencing the next section.

Question 1 How does Hooykaas view the possibility of repetition in history and in science? Refer to Unit 14, Section 3.2 (p. 113) and compare Hooykaas' view on the implications of history for the future with that of Richardson.

According to Hooykaas repetition is not possible in history because men are conscious of what has happened in the past and so are influenced by it. Even in science though experiments can be reproduced only as external factors bearing on the experiments are reproduced, but complete isolation of these factors and so identical reproduction of the experiment are not possible. Richardson says that we never learn from history whereas Hooykaas points out that because we are influenced by the past history is non-repeatable.

Question 2A Before attempting the questions based on this section you will find it useful to note the other areas of human knowledge in which the eighteenth century showed a new historical awareness.

1 Cosmology—Television Programme 6 (see also The Open University (1973) AST 281 *Science and the Rise of Technology since 1800*, Unit 2 *From Natural Philosopher to Scientist*. Note that according to Laplace the universe has a history and that the nebular hypothesis entails a time scale).
2 Physical geology—Television Programme 7, Unit 11 *Genesis and Geology*.
3 Palaeontology—Television Programme 8.
4 Biology—Unit 12 *Precursors of Darwin?*
5 Theology—Unit 14 *Religious Thought and the Idea of Evolution*.

Question 2B How did the idea of the 'historicization of nature' fit in with the Hebrew conception (see Unit 14)? Does it justify the idea of the history of mankind as merely the extension of the history of nature?

The Hebrew idea of history postulated a beginning as well as an end and required that events had to be unique to be historical (in the sense that 'history' is used in this unit), in contrast to the cyclical Greek view. The 'historicization of nature' fitted in with the Hebrew linear concept but this did not, and does not, mean that it is justified in scholarly terms. Indeed, as the unit sets out to show, a great deal of misplaced historical thinking has arisen from the assumption that patterns from the history of nature apply to the history of mankind; whereas, in fact, they may be essentially opposed.

Question 2C Contrast the principles which distinguish the history of nature and the history of mankind (compare your answer to that in Question 1).

The most important point to grasp at this stage of the arguments is that their methods are different and that confusion has arisen when the same methods have been applied to both. The history of nature takes physics as its model but whereas the laws established for physics can be demonstrated by repeatable experiments, the laws derived for the history of nature are not capable of demonstration by repeating past events. Events in the history of mankind are, similarly, not repeatable; above all the history of mankind is the re-enactment of the past in the mind of the historian. Further the knowledge of past history influences the way men act and so the history they live.

Question 3 Refer to Unit 11 *Genesis and Geology*, Section 4.3 (pp. 76–9); how do 'uniformitarianism' and 'actualism' relate to each other? Why is 'uniformitarianism' referred to as a-historical?

Actualism is a method of interpretation based on the principle that physical causes which operated in the past can be observed today: *uniformitarianism* is a system resulting from the application of extreme actualistic method, extreme in the sense that it requires that the physical causes operating today are those which operated in the past *and* that they operated at the same intensity. Hutton and Lyell both took extreme uniformitarian positions; Hutton's view is described as 'a-historical' because it rested on a system which essentially continued unchanged. Lyell's views (see Unit 11 and *Gillispie*, chapter 5) were basically similar and may go some way to explain why it took him so long to come round to even grudging acceptance of the idea of organic evolution. It is 'a-historical' because it provides either for a 'steady-state' situation or a steady rate of development (the subtle change forced on uniformitarianism to accommodate Darwinian ideas).

Question 4 Contrast the way the uniformitarian Lyell and the progressionist Miller explained the position of mankind in creation. Which earlier views can Lyell's position be related to?

The Christian theistic view of Miller saw the creation and development both of the universe and of mankind as part of the continuance of thought and will of God. Lyell's view was essentially deist (as, indeed, was Hutton's) though in superimposing the moral order of mankind on the universe, it might be more accurate to describe him as a semi-deist (see Unit 11, pp. 69–70) provided this would not commit him to a supernatural explanation for the uniqueness of man (see also Unit 12, p. 38). Recent writers have emphasized the importance in Lyell's thought of the dignity and status of man, regarding his respect for man's 'high genealogy' as a main motivating force in his opposition to Darwinian evolution. (See, for example, M. Bartholomew, *Brit. J. Hist. Sci.*, 1973, **6**, 261 and compare *Gillispie*, p. 131.)

Question 5 What compromises had to be made in the application of uniformitarianism? Did they affect it as a system or as a method?

The first compromise concerned the 'present' in which the forces which had operated in the past were held to be still in evidence and to be working at the same intensity. The 'present' was extended back 'as far as human recollection'. A particular force, e.g. volcanic eruption, might not be operating continually but had on occasion been evident during the 'present'. This allows at least some acceptance of 'cycles' in the history of nature and the compromise here is that some concession can be made to 'catastrophism' within a uniformitarian framework. The second compromise concerned the primaeval epochs when different physical conditions must have meant that basic forces operated differently from today.

These compromises and modifications seem effectively to negate the idea of strict uniformitarianism as a *system*, and drew attention to the limitations of the actualistic method.

Question 6A In what ways did Darwinian evolution demand modifications to strict uniformitarianism?

Firstly, it was historical, it posited a beginning, if not an ultimate beginning, and a progression by a non-repeatable natural selection, thus going against the idea of a world system in a steady state. Secondly, biological evolutionists gave only limited weight to geological considerations, e.g. in their discussion of the distribution of species, their isolation, etc.; organic change was not *necessarily* discussed within a geological framework. Finally, and of greatest importance, uniformitarianism came to be viewed as a uniformity of process (of 'coming-to-be' as Hooykaas puts it) rather than as a steady state (of 'being'). This meant that uniformitarianism had returned to being a methodological approach rather than a system.

Question 6B This is something of an advance warning of a question, rather than a question itself! On the basis of the exposition thus far, can you identify basic differences of approach in the expositions of the two authors (note also p. 21 of this unit)?

Hooykaas is setting the theory of evolution against the background of internal developments in the history of science, though in a less specific way than in Unit 13, particularly in its relationship to the uniformitarian approach to geology. In contrast Richardson places it in the context of the historical revolution at the beginning of the nineteenth century. However on p. 21 Hooykaas writes of the 'borrowing' by biology of a certain view from history, though insisting that the debt was mutual. Perhaps the crucial difference is that Unit 14 stresses the importance of specific theological views in the nineteenth century. This difference of approach is important and one that will be developed in later questions. A question to bear in mind is this—are they two mutually exclusive approaches, or are they simply views from different angles?

Question 7 To what extent does history repeat itself in respect to man and to nature? Can the Biblical account be called uniformitarian in its approach? (See II Peter 3:3–10.)

In the actualist sense the history of nature repeats itself continually in that the same forces continue to operate even though events resulting from these basic forces were historic because they were unrepeatable. For mankind fundamental attitudes remain unaltered but man acts consciously and is influenced by his knowledge of the past.

While the Biblical account presents certain unchanging absolutes, e.g. the nature of God and the plight of man, it claims to be a record of God's dealings with man. II Peter 3:4 appears to attack a narrowly 'uniformitarian' view of history and to advocate a 'catastrophist' position, e.g. verse 6. However, perhaps we should hesitate to apply the scientific category 'uniformitarian' to the Biblical record. Some authors have attributed uniformitarianism to the scoffers of verse 3!

Question 8 According to both Hooykaas (p. 21) and Richardson (Unit 14, p. 114) human history acquired a 'scientific' element in the nineteenth century. In what aspects do their views appear to differ?

Unit 14 relates to the revolution in historical thinking early in the nineteenth century and refers to the attempts to achieve a more objective, dispassionate analysis. In that sense history became more 'scientific'. The present unit is more concerned with the specific, and later, impact of evolution, where Darwinism was 'carried over' into social history. Moreover, the previous unit is concerned with the reverse interaction, in which a science (biology) is to some degree conditioned by an attitude to history.

Question 9 How, in Section 4.1, does Hooykaas effectively attack a whiggish historiography?

He does so by a consideration of the concepts of 'progress' and 'improvement', assessing the dangers of carrying over these ideas into the historiography of science; note particularly the paragraph on 'precursors' (p. 22).

Question 10 At the end of this section there are two important questions to be answered:

1 **To what other areas has the Darwinian idea of continuous progress been applied? What assumptions had to be made? What criticisms can be levelled at this extension of the principle?**

2 **Comment on Richardson's statement, Unit 14, p. 124, 'many leading scholars in the Church felt even the Scriptures themselves had "evolved" over long ages'.**

---

1 The idea of continuous development and the actualistic principle underlying it were widely applied to archaeology, ethnography and other aspects of prehistory. Hooykaas argues that the acceptance of the principle of continuous progress determined how the evidence was interpreted to provide evidence for the principle of continuous progress. . . . In other words the principle was transferred to other areas of investigation on the unsupported assumption that it applied, the evidence as it was found was arranged in the pattern dictated by the principle and so was shown to support the principle! The example of the tools of prehistoric man is the most telling one used—their arrangement in ascending order was made on the assumption that man must have progressed, on the basis of an actualistic reconstruction of how these tools would have been developed by modern man in primitive conditions.

2 The 'leading scholars' to whom Richardson refers, who held that the Scriptures had evolved over a great period of time, were following the same method as the archaeologists and ethnologists whom Hooykaas criticizes. Having accepted from Darwin's theory of biological evolution the idea of continuous progress they applied it to the Scriptures, using *external* forms into which they forced Scripture.

Question 11 Refer back to the preliminary question in the Introduction. To what extent is Part One of this unit applicable to the question from Unit 14?

The whole tenor of Part One of this unit is directed against the idea that the principle of evolutionary progress can be applied outside the sphere of biology. The question from Unit 14 is framed on the assumption that the 'world' (presumably in its physical, geological and organic structures) has evolved, that not only man but his ideas have evolved. Thus the two units could be perceived as dealing with essentially different things; evolutionary progress (Part One of this unit) and evolution itself (Unit 14).

Acknowledgements

Grateful acknowledgement is made to the following sources for material used in this unit:

Figures 1 and 4: Mansell Collection; *Figure 2:* Bettmann Archive.

Unit 16
The End of an Era?

Prepared by Colin A. Russell for the Course Team

ILLUSTRATION OVERLEAF

Darwin in his last years (Radio Times Hulton Picture Library).

Contents

1 Introduction

This final unit of the course gives us an opportunity to look back and briefly to take stock of the position towards the end of the last century. In particular it enables us to raise certain important questions, even if the answers to them must be partial and incomplete. For example: how much truth is there in the common belief that Darwinism dealt a death blow to orthodox Christian theology? How far was his work effectively the end of an era for religion? To what extent did it promote a revolution in scientific thinking? It is desirable that these matters should be ventilated for two reasons. Since the course deals with the period from Copernicus to Darwin (and not later) one might draw the conclusion that Darwin's work was particularly final, and that after him there was little of significance for science and belief. But a more important reason is just that a great deal of solid historical research has appeared in the last ten years or so and such work is always likely to put traditional views at risk. Moreover Darwinism still appears to be a live issue for some of those who would thrash Christian beliefs on one hand or passionately defend them on the other; representatives of both camps generate their own kind of literature and continue to do so right up to the present day. Rarely, however, do their polemics pay much attention to historical scholarship and it is no bad thing to attempt a critical reassessment of the place of Darwinism in human thought.

This unit is fairly short and by no means all of the material will be new. We shall try to pull together threads from earlier units, especially Units 12, 13 and 14. There are no broadcasts specifically associated with the unit, but television programme 11 and the last three radio programmes have strongly related themes. In the set books the appropriate passages are document 26 from *Anthology I* and chapter 14 from *Anthology II*.

We shall be looking at these two passages more fully later on. But to help you to focus on the nature of the problem it is suggested that you read the first two and the last two paragraphs of Professor Chadwick's account in *Anthology II* (pp. 282–3 and 292–3) and the first five and last four paragraphs of Huxley's account (*Anthology I*, pp. 457–9 and 480–2), bearing the following questions in mind:

1 How widespread was the acceptance of Darwinism by the 1880s?
2 How far was Darwinism a major determinant in human thought?
3 Do you notice any peculiar features of Huxley's historiography?

On point (1) both documents appear to agree on the widespread, though not universal, acceptance of Darwinism by the 1880s. Chadwick, however, is careful to point out the exceptions, and they were more numerous than could be inferred from Huxley's account. We shall see more of this later, but it is worth noting now the danger of equating religious belief with the views of the Anglican clergy—or even a section of them. Huxley's optimism needs to be qualified; it was too socially restricted, and ignored the strength of 'grass-roots' opinion in the nonconformist and Anglican circles.

Huxley was in no doubt about the seminal importance of Darwin's work, twice comparing him with Newton; and both writers agree that, for the younger generation at least, the conflict was largely over. As we shall see the later historian is much more cautious in his assessment of Darwinism as the main determinant in the intellectual trends of the later nineteenth century. It is perhaps on this point that Huxley is weakest, ignoring the pre-Darwinian changes in men's view of history, for example. By a curious inversion he even contrives to see the emergence of a theory of evolution as facilitated by the triumph of Greek ideas over Christian ones! As you will have seen from Unit 14, precisely the opposite view can be argued; cf. also *Hooykaas* (SET BOOK).

Objectives

The objectives of this unit are as follows. You should be able to:

1 Apply the general principles laid down in preceding units to the situation created by Darwinism. Amongst these are:
   (a) Applicability of theories of 'accommodation';
   (b) Inherent problems in any argument from design;
   (c) The distinction between a world picture and a world view;
   (d) The non-homogeneity of theological responses to science and the factors underlying this variation;
   (e) The importance of non-scientific beliefs in the formulation and progress of scientific doctrines;

2 Argue the case for and against the view that Darwinism closed an era of intellectual history.

3 Identify the specific nature of the challenge offered by Darwin to religious belief.

4 Differentiate between the world picture and the world view associated with evolutionary thought.

5 Assess critically the presuppositions and implications of 'social Darwinism'.

2 The Specific Challenge of Darwinism to Religion

There can be no denying that Darwin's views did constitute a formidable challenge to the orthodox religion of his time. This has already been discussed, e.g. Unit 13, Section 10. Partly as a revision for that material, and partly to draw several other threads together, you are invited to spell out as precisely as possible the exact nature of that challenge. To help in this there follow five quotations, in each of which the author puts his finger on one aspect of the confrontation.

A

If we have all, men and monkeys, oysters and eagles, developed from an original germ, then is St Paul's grand deliverance—'all flesh is not the same flesh; there is one kind of flesh of men, another of beasts, another of fishes, and another of birds'—untrue. (*Cited by A. D. White*)

B

The old argument from design in nature, as given by Paley, which formerly seemed to me so conclusive, fails, now that the law of Natural Selection has been discovered. We can no longer argue that, for instance, the beautiful hinge of a bivalve shell must have been made by an intelligent being, like the hinge of a door by man. There seems to be no more design in the variability of organic beings, and in the action of Natural Selection, than in the course which the wind blows. (*Charles Darwin*)

C

Natural laws are said to be to God what the chisel and brush are to the artist. Then God is as much the author of species as the sculptor or painter is the author of the product of his skill. This is a theistic doctrine. That, however, is not Darwin's doctrine. His theory is that hundreds or thousands of millions of years ago God called a living germ, or living germs, into existence, and that since that time God has no more to do with the universe than if He did not exist. This is atheism to all intents and purposes, because it leaves the soul as entirely without God, without a Father, Helper, or Ruler, as the doctrine of Epicurus or of Comte. Darwin, moreover, obliterates all the evidences of the being of God in the world. He refers to physical causes what all theists believe to be due to the operation of the divine mind. (*Charles Hodge, 1872*)

D

If the geologist dwelling exclusively on one class of facts, which might be paralleled by the existing creation [arrives] at conclusions derogating from the elevated position previously assigned by him to Man, if he blends him inseparably with the inferior animals

& considers him as belonging to the earth solely, & as doomed to pass away like them & have no farther any relation to the living world, he may feel dissatisfied with his labours & doubt whether he would not have been happier had he never entered upon them & whether he ought to impart the result to others. (*Charles Lyell, 1858*)

E

'Tis the crown and glory of organic science that *it does*, through *final cause*, link material to moral . . . . You have ignored this link; and, if I do not mistake your meaning, you have done your best in one or two pregnant cases to break it. Were it possible (which, thank God, it is not) to break it, humanity, in my mind, would suffer a damage that might brutalize it, and sink the human race into a lower grade of degradation than any into which it has fallen since its written records tell us of its history. (*Sedgwick to Darwin, 1859*)

What underlying difficulties does each passage recall?

From these passages one could identify the following difficulties; there were challenges offered to:

A a strictly literalistic interpretation of the Bible;
B the validity and utility of the argument from design;
C the employment of final causes in biological explanation;
D the traditional view of the uniqueness of man;
E the stability of society itself.

Undoubtedly there were other challenges as well, though not necessarily perceived as such at the time. The obvious question that presents itself is the extent to which these challenges were final. Did they in fact administer the *coup de grâce* to orthodox religion, or to some elements in it?

Before trying to suggest an answer (Section 3) it would be as well to look again at the points that were raised.

2.1 Literalistic interpretation of Scripture

This is still for some people the most obvious aspect of the Darwinian challenge. Rather than discussing it in general terms it might be helpful to spell out precisely what that issue was.

Try to read Genesis, chapters 1 and 2 from the point of view of someone freshly confronted with the Darwinian theory of evolution by natural selection. What points of conflict would you see? How serious would they be?

This depends to a considerable extent on your powers of historical reconstruction and imaginative insight. Perhaps the following are the more obvious elements in an answer:

1 The time scale: just four days required for the whole of the organic creation.
2 A specific order of events: land vegetation, marine life, land animals, man.
3 Creation of man *from the dust*.
4 Special creation of woman.
5 No hint of relationships between species, descent, modification, natural selection, etc.

In the light of events described in earlier units we can see that (1) was no new problem posed by Darwinism and not likely to be specially worrying. One could simply regard 'day' as an immense period of time. Objections (3) and (5) are both effectively based on arguments from silence; the evolutionist could accept man's creation *ultimately* from the dust of the earth, since his body undoubtedly contains the same elements. Since no mechanism was described none was proscribed—Darwin's included. It was rather like the arguments for or against the plurality of worlds. On the other hand some felt that Genesis really did give sufficient information to imply something rather definite about the process. However (2) and (4) could not be evaded in this way. But the difficulties remained only if one insisted on strictly literal interpretations; to regard the Genesis account in terms of 'accommodation' is to remove all these difficulties. These points were clear to some but not to others at the time. Perhaps the value of 'accommodation' is seen more by hindsight; the pressing question in a contemporary situation is always *where* to apply the method.

To deny to the Genesis account a literal interpretation is to lay open the *possibility* of similarly treating the New Testament narratives. In Darwin's day this could have meant a denial of the resurrection but in that case it would have evoked a sympathetic response from certain groups anxious to eliminate such claims on other grounds; such a response would have been absent—or at least less vocal—in previous times, cf. Unit 11, pp. 72–7. Nevertheless, in terms of literal interpretation it is hard to see that there were any substantial new problems raised by Darwin. In fact, as may be recalled from Units 12 and 13, there were several other areas in which problems were created, not least the apparent demise of the argument from design.

2.2 The argument from design

The words by Darwin in quotation B hit the nail right on the head. He saw, even if some of his followers did not, that the concept of natural selection could replace the idea of conscious design.

This may seem a relatively trivial matter until one recaptures in some measure the sense of *utter dependence* of many people upon arguments like those of Paley, not only to confound the unbeliever and convert the heathen, but also to bolster up the confidence of the faithful. Several other traditional arguments had taken hard knocks about this time and in a very real sense the argument from design was the last *traditional* weapon in the armoury of the orthodox. (As we shall see in Section 3 their anxiety was partly ill-founded, but we now have the advantage of hindsight.) We can sharpen this point by the following question.

What other arguments of traditional apologetics were being deprived of their force around the 1860s?

One pair of related arguments depended upon the fulfilment of prophecy and the existence of miracles. Both had been severely handled by Hume, as had the argument from design, but both displayed more resilience than perhaps Dillenberger (*Anthology II*, pp. 175–82) would allow. Indeed, numerous writers today continue to use these arguments, although in a modified form. The growing importance of Biblical archaeology from the early nineteenth century had thrown up a great deal of empirical data which were grist to the mill of those who argued the truth of Scripture from the fulfilment of its prophecies. The early Victorians were particularly impressed, for example, by the dramatic discoveries of the ruined cities of the Edomites upon which the Divine judgement had been foretold (see pp. 44-5).

Figure 1 Sir Flinders Petrie (Mansell Collection).

Yet not all were convinced, and for one author, at least, the explanation of his mixed reception was clear. In the words of Alexander Keith (1847):

> It is easy to exclude the noontide light by closing the eyes; and it is easy to resist the clearest truth by hardening the heart against it. And while, on the other hand, there are minds (and Newton's was among the number) which are differently affected by the Evidence of Prophecy, and which cannot be callous, when touched by the concentrated rays of light from heaven, whence can this great dissimilarity of sentiment arise from the same identical and abundant proof? And into what else *can* the want of conviction be resolved than into the Scriptural solution of the difficulty—an evil heart of unbelief?

Although coming perilously near to a circular argument this author offers an illuminating comment on the dilemma in which apologetics had found itself. Much the same was also true of the arguments based upon miracles. Professor Chadwick (*Anthology II*, pp. 289–90) underscores this point, emphasizing the contributions of 'history as much as science to this change of thought'. Then, thirdly, there was the array of arguments based upon the attempted correlation between geological discoveries and the Scriptural record as indicated in the very title of Buckland's *Vindiciae Geologicae*. As we have seen in connection with the Flood (Television Programme 7), such efforts ran into severe difficulties.

But despite all these minor problems the argument from design remained with its strength little abated by the assaults of Hume upon it. However, with natural selection the last prop underpinning Christian theology seemed to have been ruthlessly kicked away.

Indeed it was often asserted that teleology received its death blow from Darwinism. Consider these words of Huxley (1870):

> According to Teleology, each organism is like a rifle bullet fired straight at a mark; according to Darwin, organisms are like grapeshot of which one hits something and the rest fall wide.
>
> For the teleologist an organism exists because it was made for the conditions in which it was found; for the Darwinian an organism exists because, out of many of its kind, it is the only one which has been able to persist in the conditions in which it is found.
>
> Teleology implies that the organs of every organism are perfect and cannot be improved; the Darwinian simply affirms that they were well enough to enable the organism to hold its own against such competitors as it has met with, but admit the possibility of indefinite improvement.

Figure 2 Drawing from life of T. H. Huxley (University of Cambridge Department of Geology. Photo Edward Leigh).

That, however, is only half the story and it is interesting to compare this quotation with an extract from his essay in *Anthology II*, pp. 201–2.

Summarize the gist of Huxley's argument in this passage.

Huxley is postulating a set of teleologically conceived laws operating upon the particles of matter at a sub-biological level of organization.

This point is of great importance and has been put much more clearly by numerous modern authors, e.g.:

> Natural selection is permissive rather than constructive. It might be that the possession of an eye would make the reproductive survival of a species more probable. But . . . the possible form of an organ of vision is severely restricted by the properties of matter and the principles of engineering construction peculiar to this Universe. There are so few ways in which an eye can be made. The principle of the photographic camera and of our own eyes is one of these ways. The material parts of both men and machines can only be constructed on certain engineering principles; and only limited kinds of material are available to make them. Can we see design in the properties of the materials available for the construction of organisms? These properties are not the result of natural selection. (*C. F. A. Pantin, 1959*)
>
> If, then, we are prepared, in order to save the deterministic hypothesis, to say that system *must* have been capable of producing the evolution, we shall have to say (if we approve of the Universe!) that it was very lucky that the particular set of laws which happen to operate in the Universe was just this set. Those of us, who, on other grounds, see the Universe with its laws as the expression of Mind, will see, in the fact that the laws of physics which actually operate are capable of producing the laws of biology, only another evidence of design. (*F. H. Cleobury, 1967*)

Figure 3 An example of the use of fulfilled prophecy in Christian apologetics. The ruined city of Petra has often been cited as a dramatic example of judgements predicted in the Old Testament (e.g. Jeremiah 49:16, Malachi 1:4). To the early Victorians these discoveries came with new and welcome force (see p. 42).

EVIDENCE

OF THE

TRUTH OF THE CHRISTIAN RELIGION,

DERIVED FROM THE LITERAL

FULFILMENT OF PROPHECY;

PARTICULARLY AS ILLUSTRATED BY THE

HISTORY OF THE JEWS, AND BY THE DISCOVERIES OF RECENT TRAVELLERS.

BY

ALEXANDER KEITH, D.D.,

AUTHOR OF "THE SIGNS OF THE TIMES," AND "DEMONSTRATION OF THE TRUTH OF THE CHRISTIAN RELIGION," ETC.

THIRTY-FIFTH EDITION, STEREOTYPED.

(Third Impression.)

Opinionum commenta dies delet, Naturæ judicia confirmat.—*Cic. de Nat. Deo.*

EDINBURGH:

WILLIAM WHYTE & CO.,

BOOKSELLERS TO THE QUEEN DOWAGER.

LONGMAN AND CO., J. DUNCAN, WHITTAKER AND CO., SIMPKIN AND CO., HAMILTON AND CO., J. HATCHARD AND SON, AND NISBET AND CO., LONDON. W. CURRY, JUN., AND CO., DUBLIN. W. COLLINS, GLASGOW.

MDCCCXLVII.

Figure 3(a) Title page of Fulfilment of Prophecy, *by Alexander Keith, 1847.*

Figure 3(b) Frontispiece of Keith (op. cit.), *showing a general view of Petra from the north east.*

Figure 3(c) The ruins of a temple at Petra, from Keith (op. cit.).

Figure 3(d) A Corinthian tomb at Petra, from Keith (op. cit.).

Figure 3(e) A tomb at Petra, from Keith (op. cit.).

Thus again, it became clear that the challenge offered to orthodox belief, though real, is limited. Historically the importance is the blow dealt to traditional natural theology though it was many years before Paley's *Natural Theology* was finally withdrawn as a set book for certain examinations at Cambridge. It is interesting to notice that Darwin himself retreated from his extreme views on teleology towards the end of his life. In 1862 he had published an important book on the fertilization of orchids, and this had been widely taken as an interesting example of natural theology in the sense that it appeared to declare something of a belief in design in the universe, using such terms as 'purpose', and 'beautiful contrivance', and stressing that because useful varieties are preserved in natural selection, uses for all parts of an organism are a legitimate and necessary subject for research. In private conversation with the Duke of Argyll many years later the case of orchid fertilization was mentioned as an example of purposive design and Darwin replied, 'Well, that often comes over me with overwhelming force; but at other times it seems to go away'. At very least the remark indicates a certain ambivalence in Darwin's attitude.

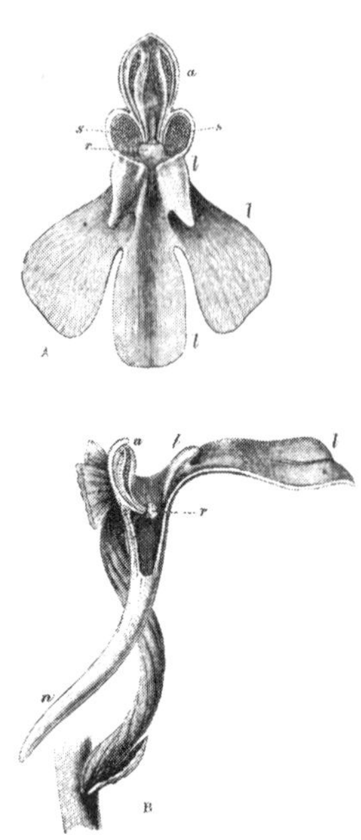

Figures 4(a) and (b) Front and side views of the Orchis Pyramidalis, *with all sepals and petals removed, except the labellum.*

2.3 The elimination of final causes from biology

Passage C, p. 40 comes from an influential theological writer of the 1870s, Charles Hodge. His objection is clearly related to this point and draws attention to one of the most serious and far-reaching aspects of the Darwinian challenge. Read the passage again and answer the following question:

Identify both old and new elements in the Darwinian challenge as seen by Hodge.

This kind of challenge was no new thing; it had all happened before when the mechanistic universe of Newton was relieved of the necessity of 'special' divine interventions. Hodge's 'atheism to all intents and purposes' is the deism of the seventeenth and eighteenth centuries. What is new is the extension of these ideas to living creatures, and the challenge is that of complete reductionism, reducing *all* phenomena, biological included, to the laws of physics and chemistry.

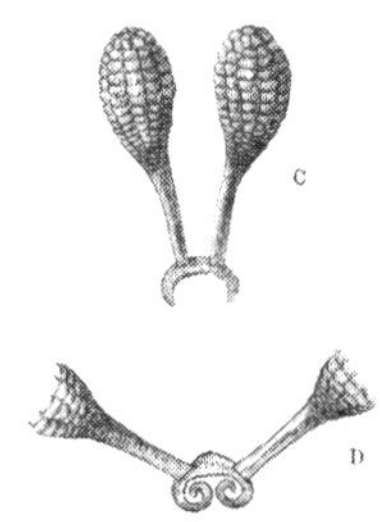

Figures 4(c) and (d) The pollen grain before (c) and after (d) contraction (all from Charles Darwin, On the various contrivances by which British and Foreign Orchids are fertilized by insects, *1862).*

But what were these laws? Laws of nature had been considered, as Hodge observed, as the instruments of God's activity, but the thrust of his objection is that Darwin has evacuated them of all theological content.

What reactions was this view likely to engender?

Certainly these ideas coincided with a good deal of popular thinking. As Mr Gladstone is said to have observed: 'Upon the grounds of what is termed evolution God is relieved of the labour of creation; in the name of unchangeable laws he is discharged from governing the world.'

More significant, perhaps, was the reaction of the American botanist Asa Gray who largely accepted Darwin's teaching. Gray was far from being an atheist (cf. Unit 13) but wrote in 1876:

> It seems to me, therefore, perfectly evident that the substitution of natural selection, by necessity, for design in the formation of the organic world, is a step decidedly atheistical. It is in vain to say that Darwin takes the creation of organic life, in its simplest forms, to have been the work of the deity. In giving up design in these highest and most complex forms of organization, which have always been relied upon as the crowning proof of the existence of an intelligent Creator, without whose intellectual power they could not have been brought into being, he takes a most decided step to banish the belief in the intelligent action of God from the organic world. The lower organisms will go next.

Figure 5 Asa Gray, photograph taken in the early 1880s (Bettmann Archive).

As might be expected, there were others who swore that Hodge's antithesis between 'physical causes' and 'operations of the divine in the mind' was an artefact of his own. His insistence that natural law had to be either *one* or the *other* overlooked the possibility that it could be both. Charles Kingsley recognized this, saying, 'It is just as noble to believe He [God] created forms capable of self-development as to believe He required fresh acts of intervention to fill the gaps He Himself had made.' Kingsley even alluded to this in a passage about Mother Carey in the *Water Babies* (my italics):

> There was once, for instance, a fairy who was so clever that she found out how to make butterflies. I don't mean sham ones, no; but real, live ones, which could fly, and eat, and lay eggs, and do everything that they ought; and she was so proud of her skill that she went flying straight off to the North Pole to boast to Mother Carey how she could make butterflies.
>
> But Mother Carey laughed.
>
> 'Know, silly child,' she said, 'that anyone can make things if they will take time and trouble enough; *but it is not everyone who, like me, can make things make themselves*'.

Figure 6 The Rev Charles Kingsley, from the photograph by Elliott and Fry (Mary Evans Picture Library).

The opposite of this view of biology is that of special creation; as Aubrey Moore observed, 'a theory of occasional intervention implies as its correlative a theory of ordinary absence'—a position which one might describe as that of semi-deism (Unit 11). Certainly some adopted this position.

One factor which again became crucial in both sides of the debate over final causes was the concept of natural law, as a good deal of contemporary writing testifies, e.g. *The Reign of Law* by the Duke of Argyll, which emphasized this point and also enquired whether varieties had been led and whether some species were not more beautiful than it was necessary for them to be. More remarkable was Drummond's *Natural Law in the Spiritual World*. This remarkable work sought to extend natural law from biological to spiritual life, just as its dominion had previously been increased from physical to biological science. It was this extension of the notion that constituted the present difficulty, for it now had to include in its purview not merely the humble plants and animals but the very summit of creation, man himself.

The ambiguity of the position for many people has been well summed up in the following lines:[1]

> A firemist and a planet,
> A crystal and a cell,
> A jellyfish and a saurian,
> And caves where cavemen dwell.
> Then a sense of law and beauty,
> And a face turned from the clod,
> Some call it 'Evolution'
> And others call it God.

Figure 7 George Douglas, eighth Duke of Argyll (Mansell Collection).

2.4 The status of man

In 1857 Darwin wrote, 'You asked me whether I shall discuss man. I think I shall avoid the whole subject, as it is so surrounded by prejudice.' However, perhaps prompted by Huxley's *Man's Place in Nature* (1863), he shed his inhibitions sufficiently to produce his own *The Descent of Man* in 1871. But his first instincts had been correct, and the storm that centred upon him owed its force largely to the displacement of

[1] W. H. Carruth, *Each In His Own Tongue*, in *The World's Great Religious Poetry*, The Macmillan Company, New York, 1941, p. 145.

man from his unique position. Such reactions had a high emotional, as opposed to intellectual, content, much as when some felt that man was displaced from the physical centre of the universe after Copernicus.[1] Yet many who objected did so on grounds that seemed to them, and indeed to us, to be fully or largely rational.

Figure 8 Alfred Russel Wallace (Mary Evans Picture Library).

The quotation D was from Sir Charles Lyell for whom this was the ultimate abyss from which he recoiled in horror. His reactions were far from untypical. Recent work has suggested that it was just this point that so troubled several other leading scientists at the time. They were concerned about the possible loss of dignity to man if he, too, were *merely* the product of natural selection (cf. Unit 13). A. R. Wallace was concerned with other implications of this position as well and he believed 'a superior intelligence has guided the development of man in a definite direction, and for a special purpose just as man guides the development of many animal and vegetable forms'. Holding an essentially dualistic view of man he supposed that Darwinian theory 'shows how man's body may have been developed from that of a lower animal form under the law of natural selection; but it also teaches us that we possess intellectual and moral faculties which could not have been so developed, but must have had another origin; and for this origin we can only find an adequate cause in the unseen universe of Spirit'.

Thus, as Moore pointed out, Wallace was making a false distinction of cause, invoking a 'God-of-the-gaps' type of explanation. He wrote, 'whatever be the line which Christianity draws between man and the rest of the visible Creation, it certainly does not claim man as the work of God and leave the rest to "unaided nature" '.

It is interesting to compare Charles Lyell's concern for 'the high genealogy of his species'. A recent analysis of Lyell's attitude to evolution suggests: 'but the "religious" aspect of Lyell's thought which is most prominent and which looms throughout all his work, is his concern, even mania, for the status and dignity of man' (M. Bartholemew).

It is curious that the sensitive Darwin does not seem to have been unduly disturbed by this problem. Conceivably his preoccupation with his own personal reputation

THE LION OF THE SEASON.

ALARMED FLUNKEY. "MR. G G-G-O-O-O-RILLA!"

Figure 9(a) 'The Lion of the Season', cartoon in Punch, *1881 (Mansell Collection).*

MONKEYANA.

Figure 9(b) 'Monkeyana', cartoon in Punch, *1861 (Mansell Collection).*

[1] But cf. Unit 2, p. 66 ff.

('the ambition to be esteemed by my fellow naturalists') left little inclination to be worried about his own species. But it is certain that Lyell and a good many others would have been most upset by Eddington's description of man as 'the gruesome result of an occasional failure of antiseptic precautions'!

Granted that the force of Lyell's objections was aesthetic rather than geological, it remains to ask in what respect Darwin's views on the status of man acquired a specific *odium theologicum*.

How did Darwinism specifically conflict with conventional religious views about the status of man? Read e.g. Genesis, chapters 2 and 3.

Some of these points we have already looked into, e.g. the Biblical claim to the uniqueness of man and the special creation of woman. One might be tempted to emphasize the clear implication that all mankind descended from a single pair of individuals. In fact, however, this was agreed on biological grounds by Darwin, Huxley and many of their followers, and (chronology apart) was hardly a problem. Moreover, the poetic form of Genesis 3 relieved many of the difficulties associated with a serpent that talked volubly and walked vertically. However it was widely held that evolution had discredited the fall of Adam or, alternatively, was discredited by it.

It was however at a much more profound level that evolution and religion came into serious conflict with regard to the status of man. The Biblical teaching that mankind had fallen was held to be hardly compatible with the optimistic view of much evolutionary philosophy, though we must at all times be careful to distinguish between the theory of evolution by natural selection and the much wider extrapolation of this into a general evolutionary world view (cf. Unit 15). Like this broad evolutionary philosophy Christianity offered hope for what is called today 'the human predicament'. But the basis of that hope was in Christ and not in natural selection, both for the individual and for humanity as a whole. It is amusing to recall that the Scottish scientist Henry Drummond was busily preparing his Lowell Lectures on *The Ascent of Man* at the very time his close friend the evangelist D. L. Moody was addressing vast crowds in America to the effect that such an ascent was impossible.

Thus Darwin had a twofold effect upon traditional views of the nature of man; he raised questions about the uniqueness of man's ancestry and he tended to minimize the extent of his predicament. The impact of both these points was more psychological than logical. On the first one there are plenty of hidden assumptions in the view that man's descent determines his nature, and on the second it is highly doubtful if Darwin was the only or even the most important agency. Central to the Biblical doctrine of man is the concept of sin, associated with the fall of Adam. It is not merely the twentieth century that finds such a concept distasteful and unacceptable and there were plenty in the early nineteenth century who declined to receive it. George Eliot, J. A. Froude, and others helped to establish a climate of ethical revolt against dogmatic Christianity some time before the *Origin* was published. Associated with this was a strong current of rebellion against the Biblical doctrine of punishment, manifested alike in the case of Bishop Colenso and the *Essays and Reviews*. And that culminated in the famous action of the Privy Council in clearing two of the authors of ecclesiastical charges brought against them (see Unit 14), about which one wag remarked that the Lord Chancellor 'dismissed hell with costs, and took away from orthodox members of the Church of England their last hope of everlasting damnation'. It is against this background that the reception of Darwinism must be judged.

2.5 The stability of society

This quotation E comes from Sedgwick's letter to Darwin upon his receipt of the *Origin of Species*. It denoted a deep fear that the demotion of man would be accompanied by a disintegration of society. It was far from being a hysterical reaction to a totally new kind of view and a reluctance to change one's viewpoint. Sedgwick could see that there had been over the Christian centuries a coherence to society that derived in large measure from the Biblical doctrine of man. If this were to be destroyed the outcome was, in his view, too terrible to contemplate. Again he was not alone in his fears. That the theory of evolution had profound effects upon man's view of himself as a social animal is undoubtedly true, and Sedgwick's instinct was right. This came about not merely from an overthrow of traditional religious views of man and the undermining of the authority of certain parts of Christian teaching. It also came about from an extrapolation of the scientific concept to the philosophical world view and this is a topic we shall deal with in the following section.

3 The Response to the Challenge

In this section we shall be noting some of the changes in the non-scientific climate of opinion in the years after Darwin to which his work could arguably have made some contribution. The field is a vast one and it is too much to hope that this short study will enable you to make a scholarly assessment of the extent to which Darwinism was a major influence on human thought generally, but it may indicate some of the directions from which answers could be expected. We shall look at seven aspects of the subject, and begin by examining corresponding quotations. Please read them in the light of the questions that follow.

A

All opposition had availed nothing: Darwin's work and fame were secure. As men looked back over his beautiful life—simple, honest, tolerant, kindly—and thought upon his great labours in a search for truth, all the attacks faded into nothingness. There were indeed some dark spots, which as time goes on appear darker . . . . But all this opposition may be reckoned among the last expiring convulsions of the old theologic theory. Even from the new Catholic University at Washington has come an utterance in favour of the new doctrine, and in other universities in the Old World and in the New the doctrine of evolution by natural selection has asserted its right to full and honest consideration. More than this, it is clearly evident that the stronger men in the Church have, in these latter days, not only relinquished the struggle against science in this field, but have determined frankly and manfully to make an alliance with it.

B

Religious men believe with Agassiz that facts are sacred. They are revelations from God. Christians sacrifice to them, when duly authenticated, their most cherished convictions. That the earth moves, no religious man doubts. When Galileo made that great discovery, the Church was right in not yielding at once to the evidence of an experiment which it did not understand. But when the fact was clearly established, no man sets up his interpretation of the Bible in opposition to it. Religious men admit all the facts connected with our solar system; all the facts of geology, and of comparative anatomy, and of biology. Ought not this to satisfy scientific men? Must we also admit their explanations and inference? If we admit that the human embryo passes through various phases, must we admit that man was once a fish, then a bird, then a dog, then an ape, and finally what he now is? If we admit the similarity of structure in all vertebrates, must we admit the evolution of one from another, and all from a primordial germ? It is to be remembered that the facts are from God, the explanation from men; and the two are often as far apart as Heaven and its antipode.

C

Up to this time no word has been spoken to reconcile Christianity with Evolution, or Evolution with Christianity. And why? Because the two are one. What is Evolution? A method of creation. What is its object? To make more perfect living beings. What is Christianity? A method of creation. What is its object? To make more perfect living beings. Through what does Evolution work? Through Love. Through what does Christianity work? Through Love. Evolution and Christianity have the same Author, the same end, the same spirit. There is no rivalry between these processes. Christianity struck into the Evolutionary process with no noise or shock; it upset nothing of all that had been done; it took all the natural foundations precisely as it found them; it adopted Man's body, mind, and soul at the exact level where Organic Evolution was at work upon them; it carried on the building by slow and gradual modification; and, through processes governed by rational laws, it put the finishing touches to the Ascent of Man.

D

But if it be said, 'After all, there is no why; the doctrine of Evolution, by doing away with the theory of Creation, does away with that of final causes,' let us answer boldly, 'not in the least.' We might accept all that Mr Darwin, all that Prof. Huxley, all that other most able men have so learnedly and acutely written on physical science, and yet preserve our natural theology on the same basis as that on which Butler and Paley left it. That we should have to develop it I do not deny.

E

All honour to the peaceful conqueror who, by years of unhasting, unresting research annexed these new provinces of thought to the knowledge of man, and therefore to the glory of God! All honour to the herald and archaeologist of our race, who has unrolled in all its length and breadth the genealogy of the antiquity of man, and the antiquity of his habitation! All honour to the bold yet reverent touch which, in the Temple of the Most High, not made with hands, rent assunder from the top to the bottom the veil that concealed its full proportions, and levelled its ever-widening, ever-lengthening vistas backward into the furthest past of memory, and forward to the endless future of hope. Not the limitation, but the amplification of the idea of God, is the result of the labours of such a student. Not the descent, but the ascent of man is the final result of his speculations.

F

We most of us believe, that we, the higher races, have progressed and are progressing. If so, there must be some state of perfection, some ultimate goal, which we may never reach, but to which all true progress must bring us nearer. What is the ideally perfect social state towards which mankind ever has been, and still is tending?

G

The missionaries have done much in improving their [the Tahitians'] moral character, and still more in teaching them the arts of civilization.

Try to identify a different aspect of the Darwinian impact in each of these quotations. Can you infer anything about the authors?

The following are possible lines upon which an answer might be given:

A

This relates to the theological reception of Darwinism, and points to the polarization of religious opinion that it engendered. It is yet another and superb example of distilled Whiggery from the pen of the inimitable A. D. White (1898; compare Unit 1).

B

Here we obviously have a response from a theologian, or at least from a non-scientist ('scientific men' are 'they' not 'we'). Darwin is rejected but in a temperate way and with a good knowledge of both past and contemporary science. The author was in fact Charles Hodge (1874).

C

This unusual passage probably strikes you as a remarkable hotch-potch of science and ethical religion, with an attempt to unite evolution and Christianity. It could hardly have been written by anyone else except Henry Drummond, and comes from his *Ascent of Man* (1894). This is a different religious response which we may characterize as absorption or synthesis.

D
Here we see someone, probably a cleric, wrestling with the Darwinian assault on natural theology and trying to find a new direction for it. The writer was Charles Kingsley in the 1870s.

E
This passage sounds like an official tribute to a pioneer in geology and its language suggests a clerical rather than a scientific origin. It comes from the funeral sermon for Sir Charles Lyell, preached on 28 February 1875, in Westminster Abbey by the Dean, A. P. Stanley. Although ostensibly about 'the religious aspects of geology' it clearly reflects the influence of Darwin as well as Lyell. It represents a new movement of liberal optimism ('the endless future of hope') that clearly owed much to developments in science. Stanley was a typical Broad Churchman and represented all those Anglicans who viewed the future through spectacles with a distinctively rosy tint.

F
This quotation is political rather than theological and relates to what came to be called social Darwinism. It comes from A. R. Wallace's *Malay Archipelago* of 1890. You might have identified the author by his strong humanitarianism, combined with a broad evolutionary philosophy.

G
This is rather a catch, but the author was Charles Darwin himself. It relates to his observations of missionary work in Tahiti. At the close of this section we shall look at the 'missionary explosion' of the nineteenth century in the light of Darwinism.

Each of these seven aspects will now be examined briefly in turn.

3.1 Polarization of theological opinion

In the first extract White refers to the 'expiring convulsions of the old theologic theory.' Convulsions there certainly were as many people experienced crises of belief over the apparent conflict between Genesis and Evolution. But it is wrong to suppose that the 'Old Theory' (presumably special creation) was in its death throes in 1896. The fact is that right until the modern period opinion has been divided on the subject, although for differing reasons at different times. In chapter 14 of *Anthology II* (which you should read now) Professor Chadwick speaks of men talking 'as though the conflict was over' by 1900—four years after White. But he also points to the sense of dissatisfaction bequeathed to the twentieth century, and even today responsible theological opinion is divided, if not about the fact of evolution, at least about the manner of it and its relevance to religious belief.

It has been suggested that the nineteenth century was the greatest for theology since the fourteenth, and even if your knowledge of the subject is small you should have gained from the *Anthology* reading some impressions of the bewildering variety of opinions and schools. Quite apart from the debates associated with the Tractarian movement in the Church of England, the new Biblical criticism and the other theological upheavals (cf. Unit 14), the nineteenth century saw the emergence of groupings which reacted in very different ways to the Darwinian challenge: materialists, modernists, liberals, fundamentalists—these are some of the labels we hang round the necks of the principal participants, often with scant regard to their suitability. And yet the constraint to do so springs from the very complexity of the scene and lends credence to the view that, for Christian theology at least, Darwin did mark the end of an epoch. However, there are two aspects of the matter that need a little more attention.

In Chapter 14 of *Anthology II* what evidence can you find that (a) the polarization of attitudes to Darwinism was not resolved by the end of the century and (b) it is possible to overstate the importance of science in the religious crises?

Apart from the case of Bishop Ryle there were 'large numbers of worshippers who . . .

knew their faith rested upon moral considerations inaccessible to the physical sciences' (p. 292). You may have noted both here and in Unit 14 several references to 'educated' Christians and may also have (rightly) inferred from page 288 that Roman Catholicism officially tended to remain uncommitted to Darwinism.[1] As to the other challenges to orthodoxy it is clear that Darwinism was only one of a group of pressures upon religion, others being changing views of history and the new Biblical criticism. It is hard to overemphasize the importance of this point. R. W. Church suggested in 1861 that the row over Darwin's *Origin of Species* would have been a lot noisier had there not been a much greater row going on over *Essays and Reviews*. Again the first reports (in 1872) of the Babylonian story of the Flood known as the Gilgamesh Epic led many to jump to the conclusion that the parallels of the Genesis account implied a dependence of the latter upon earlier myths and folklore, thus again impugning its authority as a divine revelation, especially in contrast with that of science. Later scholarship rendered this view much less tenable but its compatibility with other aspects of Victorian thought made it peculiarly hard to dislodge.

Figure 10 The Rt Rev John Charles Ryle, Bishop of Liverpool, photo by J. E. Mayall (Mansell Collection).

3.2 The rejection of Darwinism

Huxley's article (26 in *Anthology I*) on the reception of Darwinism offers some revealing insights not only into the matter ostensibly under discussion but also into Huxley's own attitudes to it. Please read the rest of the article now and then answer the following questions.

1 With what modern sociological phenomenon does Huxley associate a change of viewpoint of the time he was writing?
2 Why was Darwinism so attractive to Huxley?
3 There is a splendid piece of unconscious Whiggery. Can you identify it?
4 What obstacles to the reception of Darwinism does he identify? Does he fully appreciate their nature?

The following are some of the points that could have emerged in your answers:

1 Huxley was acutely aware of the 'generation gap' (pp. 457, 459–60, 482). Despite his identification of himself as a contemporary of Darwin ('us ancients', etc.) he makes it clear that *his* is the modern attitude, and so aligns himself with the majority of his readers. It is an old literary ploy, but by no means inappropriate in this case. Weismann described what happened in Germany: 'Darwin's book fell like a bolt from the blue; it was eagerly devoured, and while it excited in the minds of the younger students delight and enthusiasm, it aroused among the older naturalists anything from cool aversion to violent opposition.' And in England the saga of the Gosses, *Father and Son*, is a poignant case of the same thing.

2 It seems that Huxley was deeply motivated by the laudable desire to be guided by empirical facts rather than a received authority (pp. 465–6). But then, who would admit the contrary? We return to this point later. There is no doubt from his other writing that Huxley's attitude owed much to a deep-rooted anti-clericalism which, combined with a flair for pungent polemics, inspired him to attack the religious establishment whenever possible. When the time came his role as 'Darwin's bulldog' was as predictable as it was effective. Darwin, it has been argued by some, gave Huxley the opportunity he required for his assaults.

3 In several places Huxley shows signs of a Whiggish approach to science in a typical Victorian manner, e.g. in the first paragraph on page 457. But the best example is probably the passage on page 480 when of Paley it is said, 'he proleptically accepted the modern doctrine of Evolution'. However, to be fair, these examples must be set against the percipient footnote on page 467.

4 Huxley identifies the 'deification' of 'chance', the harm to teleology, and the anti-theistic tendency of evolution. He seriously underestimates the difficulties

[1] There were individual exceptions, e.g. Mivert (see Unit 13).

facing those believers required to reinterpret parts of their Bible in non-literalistic terms; these were related to the difficulties over 'accommodation'. The authority of the Scriptures was a prime article of faith and, for many, seemed to be taking a hard knock if science was to be excluded from that authority. It is scarcely surprising that many Christians without the benefit of much formal education found the shift in viewpoint just too big to be accomplished. Again, Huxley does not appear to be sensitive to the other problems associated with a changing status for man.

Despite his anxiety to identify ecclesiasticism with anti-scientific obscuranticism, Huxley would probably have been surprised by the persistence of anti-evolutionary views in the Church, right up to our own day. Maybe he would have been more surprised by the survival of the Church. These views were not confined to certain extremist sects ('hyper-orthodox' is a better term than 'fundamentalist'), and to gain some insight into the strength of this persistent tradition we shall look briefly at one Victorian theologian, Charles Hodge, who, as we have seen, was unwilling or unable to accept Darwinism.

Hodge was, in the words of John Dillenberger, 'a responsible, urbane and learned conservative whose position contrasts sharply with its fundamentalist derivatives . . . , the most influential theologian in the Presbyterian church of his time and the star of Princeton Theological Seminary'. Hodge did not oppose Darwinism on the grounds of narrow literalism. His attitude is given in extract B on page 50.

Why was Hodge concerned to oppose Darwinism?

Ostensibly, the passage suggests, this was because of the distinction between facts and theories. He regarded the evolutionary hypothesis as a theory with a fundamentally different epistemological status from, say, the theory of Copernicus or of Galileo. And, of course, he was right. It was a theory about the past—about history—and its verification is nothing like as simple a matter as Huxley made out. This point is further discussed in Radio Programme 9 and Unit 15. Hodge's refusal to be rushed into acceptance of a hypothetical scheme of this nature was, of course, powerfully reinforced by the difficulties mentioned in Section 2, or at least by some of them.

The passage also hints at another objection which was, in fact, crucial for Hodge; the status of man: 'man is not a developed ape'. This problem, associated with the apparent obliteration of God's activity in the world, led him to equate Darwinism with atheism (the quotation on page 40).

It is easy to criticize Hodge and his successors, many of whom uncritically accepted his position without the insight that informed it. But, to quote Dillenberger once again, 'the Fundamentalist [*sic*] attack on evolution obscured the genuine problem which evolutionary thought presented to the churches. But we must also add that the Fundamentalist passion was a reaction to a modernist theory which still carried the Christian name but had lost the Christian substance'.

3.3 Attempts at synthesis

If some parts of the Church were compelled to reject the Darwinian thesis others, as we have seen, reacted differently and did their best to come to terms with it. In retrospect it is quite remarkable how few there were who recognized elements of congruity between Christianity and aspects of Darwinism. To quote C. E. Raven (1956):

> It is one of the ironies of history that Christendom which by its own Scriptures was committed to belief in an ever-working God (e.g. John 5: 17), in a progressive revelation still incomplete (John 16: 13), in suffering as the characteristic of the creature (Rom. 8: 18-23), and the means to perfection (Hebr. 2: 10), and in fuller life as the divine purpose (John 10: 10) should have so signally failed to maintain this belief when faced with the challenge of Darwinism.

In Britain the most notable exception is that of Henry Drummond.

In extract C (p. 51) what aspects of Drummond's thought would have been foreign to Darwin?

The short answer must surely be: in almost every respect. Drummond confessed to see a purpose and design in nature; he assured his readers that evolution worked through 'love' rather than through a struggle for existence (cf. Unit 13); he extended Darwin's strictly physical approach to the mind and soul; he saw the whole process in terms of a specifically Christian religion; and for him the culmination of the whole process lay in the appearance and perfection of man. The title of his book, *The Ascent of Man*, stands in sharp and symbolic contrast to Darwin's own *The Descent of Man*.

Drummond had earlier received great acclaim as the writer of a best-selling devotional work, *The Greatest Thing in the World*, and for another attempt at synthesizing science with religion in *Natural Law in the Spiritual World* (1884). An amusing example of Drummond's argument can be found in his chapter on 'Parasitism' where he applies this biological concept to the eminently Victorian habit of church-going.

Figure 11 Henry Drummond, engraved from an original photograph by Lafayette (Mansell Collection).

> One of the things in the religious world which tends most strongly to induce the parasitic habit is going to church. . . . The danger of church-going largely depends on the forms of worship, but it may be affirmed that even the most perfect church affords to all worshippers a greater or less temptation to parasitism. It consists essentially in the deputy-work or deputy-worship inseparable from church or chapel ministrations. One man is set apart to prepare a certain amount of spiritual truth for the rest. . . . What is meant to be a stimulus to thought becomes a substitute for it. The hearer never really learns, he only listens. And while truth and knowledge seem to increase, life and character are left in arrear. Such truth, of course, and such knowledge, are a mere seeming. Having cost nothing, they come to nothing. The organism acquires a growing immobility, and finally exists in a state of entire intellectual helplessness and inertia. . . . Where the worship, again, is largely liturgical the danger assumes an even more serious form. . . . His senses stirred by ceremony, now soothed by music, the parasite of the pew enjoys his weekly worship—his character untouched, his will ungraced, his crude soul unquickened and unimproved.

Drummond's efforts, however, though admired by some, were trenchantly criticized by those who held extreme positions in either direction. His earlier work was assailed by one conservative writer in a book entitled *The Strangest Thing in the World*, while S. Laing, a popularizer of advanced secularism, in a long review damns the *Natural Law* as 'showing the length to which a learned and sincere man, who starts from the predetermination to believe a particular creed, can go on inventing arguments in its support, which, if they were worth anything, would really be most conclusive against it'.

Drummond, a scientist, was distinguished chiefly by the degree of imaginative insight he brought to bear on his synthesis. He stands in the same tradition as Asa Gray, the American botanist, who found no difficulty in reconciling his religious belief with much (though no all) of the Darwinian theory. The efforts of men like Gray and Drummond left an indelible mark on theological thinking of their own day and later. Chiefly, perhaps, this was in the growth of the liberal optimism that we shall turn to in a moment. First, however, there is another aspect of thought that requires a brief reference.

3.4 Changed directions for Christian apologetics

We have already seen that the argument from design received, to say the least, a severe jolt in the concept of natural selection. It had to be, and was, modified. Amongst Roman Catholics especially the old kind of arguments continued to be used (see *Anthology II*, p. 288), and may still be encountered today. But a new direction lay ahead: Charles Kingsley was aware that it was in man's understanding of the nature of God's activity that changes had to come, and had to be reflected in a new apologetic. In a recall to what has been frequently referred to in this course as Biblical theism he asserted:

> [Men] find that now they have got rid of an interfering God—a master magician as I call it—they have to choose between the absolute empire of accident, and a living, immanent, ever-working God. . . . Verily God is great, or else there is no God at all.

How, then, was this to be translated into a persuasive apologetic argument? For Drummond the answer was clear—a new appeal to the personal element in experience which tended later to be found compatible with the ideals of existential thinking in the twentieth century. Here is a colourful and not untypical example:

> Miracles; but that question is thrown at my head every second day;
> 'What do you say to a man when he says to you, "Why do you believe in miracles?"?'
> I say, 'Because I have seen them.'
> He says, 'When?'
> I say, 'Yesterday.'
> He says, 'Where?'
> 'Down such-and-such a street I saw a man who was a drunkard redeemed by the power of an unseen Christ and saved from sin. That is a miracle.' The best apologetic for Christianity is a Christian. That is a fact which the man cannot get over. There are fifty other arguments for miracles, but none so good as that you have seen them. Perhaps you are one yourself. But take you a man and show him a miracle with his own eyes. Then he will believe it.

This approach arises directly from Drummond's attempt to integrate his science with his theology. As he observed, 'The unit of physics is the atom, of biology the cell, of philosophy the man, of theology the Christian.' It is thus quite apparent that for some, at least, Darwinism had the effect of stimulating rather than retarding the exploration of Christian theology. Nor was this limited to the Protestant thinkers; J. H. Newman welcomed the *Origin* because it seemed to offer an analogy with his own concept of mental and spiritual development.

In the years that followed the widespread acceptance of biological evolution Christian apologetics were in fact to enter a new period of consolidation. Forced to examine their presuppositions much more closely than before, the defenders of Biblical theology reacted in a variety of ways which are outside the scope of this unit to examine. But it is worth recording that many have regarded the disengagement of theology from science in terms of gain rather than loss, conceiving that the truth or otherwise of propositional theology is not best served by association with specific items of the scientific creed. Such men have seen their task as rather the removal of particular objections to Christian belief (one of these being the 'conflict thesis' about its relation with the scientific enterprise), and the establishment of new links at a deeper level with many other areas of human experience, of which science is just one component.

Others, of course, continued to argue along traditional lines, regarding Darwinism as a mere episode requiring only minor adjustments to their approach. For such people a saving feature of the situation was the distinction between man's body and soul. Rejecting any mechanistic denial of the latter they believed, quite rightly, that for the New Testament writers this distinction was of prime importance. Indeed, so great was the reaction against a mechanistic conception of man that many of the later Victorian leaders of the scientific community, especially those who worked in the physical sciences (see below, p. 61), displayed great interest in the so-called spiritist phenomena and tried to incorporate them within a scientific framework. Perhaps one of the clearest statements of Christian apologetics in terms of this distinction may be found in Frederick Temple's Bampton Lecture of 1884, 'The Relations between Religion and Science'. This is the conclusion:

> As Science cannot yet assert that life was not introduced into a world that was made habitational by a direct creative act so too Science cannot yet assert, and it is tolerably certain will never assert, that the higher and added life, the spiritual factor, which is man's characteristic prerogative, was not given to man by a direct creative act as soon as the body which was to be the seat and the instruments of that spiritual factor had been sufficiently developed to receive it. . . .
>
> In conclusion, we cannot find that Science, in teaching Evolution, has yet asserted anything that is inconsistent with Revelation; in this we assume that Revelation was intended not to teach spiritual truth only, but physical truth also. Here as in all similar cases, we find that the writer of the Book of Genesis, like all the other writers in the Bible, took nature as he saw it, and expressed his teaching in language corresponding to what he saw. And the doctrine of Evolution, in so far as it has been shown to be true, does but fill out in detail the declaration that we are 'fearfully and wonderfully made; marvellous are Thy works; and that my soul knoweth right well.' There is nothing in all that science has yet taught, or is on the way to teach, which conflicts with the doctrine that we are made in the Divine Image; rulers of the Creation around us by a Divine superiority, the recipients of a Revelation from a Father in Heaven, and responsible to judgement by His law. We know not how the first human soul was made, just as we know not how any human soul has been made since; but we know that we are, in a sense in which no other creatures living with us are, the children of His special care.

3.5 Growth of liberal optimism

It could be said of both evolution and Christianity that each had a pessimistic and an optimistic side. Natural selection presents a sombre aspect by virtue of its wastage and cruelty, whereas its ultimate goal was deemed by some writers to be perfection itself. Similarly Christianity, though stressing the plight of man and of unaided human reason, was peculiarly a message of hope and of divine deliverance. So perhaps it is not surprising that the 'optimistic' elements of both should be combined together, though at the expense of losing something from each. This injection of evolutionary optimism into religious thought towards the end of the nineteenth century led to the emergence of a form of liberal theology that stressed man's upward path and the agreeable prospect for humanity. The affinities with Drummond were obvious, and the movement owed much to other theological trends, including a tendency to discard supernatural elements from theological dogmas. Thus a popular American preacher, H. W. Beecher, wrote that 'Simple religion is the unfolding of the best nature of man toward God; man has been hindered and embittered by the outrageous complexity of unbearable systems of theology.' Others in the same tradition were Lyman Abbott, and John Fiske, and this became one of the formative influences on theology in the early twentieth century.

This evolutionary theology had its own internal weaknesses and some would say it was scarcely Christianity at all. The American theologian H. R. Niebuhr reacted against it, observing:

> Evolution, growth, development, the culture of the religious life, the nurture of the kindly sentiments, the extension of humanitarian ideals, and the progress of civilization took the place of the Christian revolution . . . A God without wrath brought men without sin into a kingdom without judgement through the ministrations of a Christ without a cross.

Other writers perceived other weaknesses, and it tended to be very much over the heads of ordinary people. The heaviest blow came in the agony of one man, Karl Barth, who discovered in August of 1914 that nearly all of his respected teachers of this persuasion had subscribed to a declaration, signed by 93 German intellectuals, in support of the Kaiser's war policy. From that time on for him 'The theology of the nineteenth century had no future'. His leading position in the world of theology for the next half century and more ensured that the lesson was not forgotten. Later W. R. Inge, Dean of St Paul's, was to comment on 'the almost complete absence in historical Christianity of any idea of evolution in man, and of progress for society', remarking of Darwin's own evolutionary optimism, 'how little even the greatest men can rise above the fixed superstitions of their age!'

3.6 The emergence of social Darwinism

Figure 12 Herbert Spencer (Mansell Collection).

It was not merely upon the field of Christian theology that Darwin exerted a considerable influence. Perhaps even more important, because more lasting, was the extension of his ideas to society. One may well agree with Professor Spanner that 'To seek to give values a Darwinian basis is about as sensible as to prescribe monogamy as a social ideal because the elementary units of positive and negative electricity are equal', but the fact remains that it was done.

The leading spirit here was undoubtedly Herbert Spencer who, in 1860, published a prospectus for an enormous five-part encyclopaedia dealing with biology, psychology, sociology and morality from the point of view that behind all was a fundamental process of evolution. This view has been much derided in our own time, doubtless in part for the same reasons that led to its expulsion from theology. Denoted by Toulmin as 'Evolution with a capital E', by others as 'myth', 'dogma', the idea had received hard knocks in its own day. Thus Huxley observed, 'Let us understand, once and for all, that the ethical progress of society depends, not on imitating the cosmic process, still less in running away from it, but in combating it.'

Darwin himself, as we have seen, was not concerned with the application of his theory to social problems, but Spencer's 'grotesque paeans of progress' (as Dean Inge called them) did not fall on deaf ears. Indeed his audience was immense and he was influential in fostering both an evolutionary view of society and a social conception of evolution.

Social Darwinism had another side, however. Underlying the widespread belief in social progress lay divergences of view as to how this was to be achieved: was it to be by competition or by cooperation? Those who believed the former—and supposed survival of the fittest to apply equally to society as to nature—were not reluctant to think in those terms within early twentieth-century capitalism. One has only to recall Rockefeller's words quoted in Unit 15. In similar vein the German philosopher Nietzsche, having disposed of God, proclaimed the gospel of survival of the fittest by exalting strength and despising weakness. It is not hard to see why his works were widely acclaimed in Nazi Germany, nor why Mussolini was presented with a copy of them by Hitler. Nor is it difficult to recognize Nietzsche as the incarnation of the philosophy of despair that, like its antithesis of liberal optimism, looked to Darwin for its inspiration.

Again, the extraordinary ambivalence of social Darwinism enabled it to be used almost as well by the political left as by the right: one could deem 'natural' a struggle between classes for survival and improvement. It is well known that Karl Marx wished to dedicate *Das Kapital* to a reluctant Darwin, though he was probably nearer in sympathy to Lamarck. He wrote, 'Darwin's book is very important and serves me as a basis in natural science for the struggle in history'. For those with socialist ideals, such as A. R. Wallace, the law of the economic jungle made little sense. According to R. M. Young, 'socialism and nineteenth-century evolutionism were always very uneasy bedfellows', and in the conflict between them Wallace chose socialism. He went on to quote an anonymous commentator of 1886 who said, 'The adoption of a science as a creed is a curious feature of socialism, for science is not, we think, on the side of socialism. The scientific doctrine of today is that progress has been effected not through equality but by means of inequality.'

3.7 The expansion of Christianity

The extract from Darwin (G on p. 51) is an interesting reminder of the naturalist's interest in Christian mission, and in particular the South American Missionary Society to which he gave considerable support (cf. the close of Unit 13). It helps to underline a very curious fact, that the century which saw the arrival of the theory of evolution saw also the greatest expansion of the Christian church since the first century: 'the great century of Protestant mission', it has been called. Whether there is any connection between the two is very much an open question. Perhaps one may venture the suggestion that, for the human individual (as opposed to the race), evolution was very much a philosophy of despair, and the aggressive expansion of Protestant and Roman Catholic Christianity in the late nineteenth century was as much a reaction against it in the sphere of religion as militant socialism had been in the realm of politics. Be that as it may, there is one fact that does emerge with clarity.

The faith of individuals was sometimes badly shaken by science, and the secularization of society proceeded apace. Yet the view that during the post-Darwinian period the supposed decline of Christianity in the West was a response to pressures from scientific discovery is simply a myth. Indeed it is interesting that 1859 saw not only the publication of *The Origin of Species* but also a religious revival that swept America and Britain. Later the expansion of missionary work overseas drew heavily on those most exposed to Darwinian teaching, men of the colleges and universities of Britain and the U.S.A. One looks in vain for evidence of a failure of nerve or loss of confidence and in fact discovers just the opposite. It seems empirically true, whatever the reason, that that section of the church loosely called evangelical was largely unaffected by changes in the scientific climate of opinion.

Again it may be asked whether many of the commentators on the period have not made the mistake of concentrating on the wise and the educated, neglecting the simple facts of social division in Victorian Britain. For every one who denied evolution because it was incredible there must have been twenty who disregarded it because it was irrelevant; not irrelevant in a sophisticated theological sense but simply in that it had no obvious bearing on their lives or beliefs. In view of this we may very much doubt whether, for such people at least, Darwin did mark the end of an era. Two quotations may make the point; the first is from a London church periodical of 1872:

> The much talked about 'Natural Selection' and 'Development', so far from making hard things plainer, and natural phenomena more easily understood, carry with them such surprising difficulties as are absolutely dumfoundering (*sic*) to common minds, though even contradictions may weigh light as feathers with those great votaries of science who reign in an atmosphere never breathed by the vulgar multitude.

The second consists of the satirical words of Pooh-Bah from *The Mikado*:

> I am, in point of fact, a particularly haughty and exclusive person, of pre-Adamite ancestral descent. You will understand this when I tell you that I can trace my ancestry back to a protoplasmal, primordial atomic globule. Consequently my family pride is something inconceivable. I can't help it.

4 The Reactions of Science

4.1 The integration of science?

It has frequently been suggested that one of the great achievements of Darwin's theory was the integration of science, and indeed of a wider corpus of human knowledge. This is how Henry Drummond put it:

> Evolution involved not so much a change of opinion as a change in man's whole view of the world and of life. It is not the statement of a mathematical proposition which men are called upon to declare true or false. It is a method of looking upon Nature. Science for centuries devoted itself to the cataloguing of facts and the discovery of laws. Each worker toiled in his own little place—the geologist in his quarry, the botanist in his garden, the biologist in his laboratory, the astronomer in his observatory, the historian in his library, the archaeologist in his museum. Suddenly these workers looked up; they spoke to one another; they had each discovered a law; they whispered its name. It was Evolution. Henceforth their work was one, science was one, the world was one, and mind, which discovered the oneness, was one.

Comment on the novelty and appropriateness of Drummond's assessment of the integrative effects of evolution upon science.

The thing that shines out of this paragraph is surely the intensity of Drummond's belief that man's attitude to nature really had taken a new turn, and that 'henceforth' all was going to be very different. So undoubtedly many thought at the time, but the element of novelty can be over-stressed. Since the Greeks men have believed that knowledge was one and evolution was certainly not the first idea that purported to demonstrate that this was so. A remarkable feature of Drummond's list is that it was far from comprehensive, and excluded the physicists and the chemists.

Another way in which the weakness of Drummond's position may be exposed is by pressing the question: *how* did evolution supply the unity he declared? One may presume the answer would be that it provided a common viewpoint for all the disciplines mentioned. But what senses of the term 'evolution' could be said to be elements common to all? Clearly not natural selection or survival of the fittest in the case of astronomy and physical geology. And, as Unit 15 points out, the connection between Darwinian evolution and human history and archaeology is entirely gratuitous. In fact the *only* common element would seem to be simply *change*, and there was nothing particularly new in the realization that all the areas mentioned were subject to that! What *was* likely to be new, perhaps, was the *intensity* of that realization, and the quickened understanding of the place of human affairs against the background of biological evolution.

There is no doubt that the life sciences in general did receive a stimulus from Darwinism, though some have doubted whether it was a stimulus in the right direction. Inevitably the acceptance of a paradigm (or theoretical framework) as strong as this put severe constraints upon all the work done within it. Thus it has been suggested that early studies in embryology were hampered by the 'recapitulation theory' according to which the embryo developed through stages corresponding to different periods of the species' past evolution. Haeckel considered that the only relevant questions to ask were therefore phylogenetic ones, i.e. those relating to ancestry. Other areas of study that owed much of their impetus to Darwin were those relating to mutations, cell-division, parasitism, vestigial organs, ecology, and (later) biometrics and eugenics.

Figure 13 Ernst Haeckel (Mary Evans Picture Library).

Integration of subject-matter is one thing, integration of approach is quite another. C. E. Raven has perceived that, in one respect at least, controversy within the life sciences helped to achieve a disintegration rather than the reverse. He wrote:

> After its first outbreak in 1860 the Darwinian controversy became sharper and more destructive. As almost always happens, the struggle produced bad effects on both sides. Haeckel and Weismann exaggerated special elements in the theory, which gave it a more rigidly materialistic emphasis and eliminated from it the influence of environment and of use and disuse. T. H. Huxley, who was principally an anatomist and taxonomist and knew little of living organisms, stressed the gladiatorial picture of nature 'Red in tooth and claw with ravin' ' and prompted the cleavage between the scientists of museum and laboratory and the observers of wild life in the field.

A similar view was expressed by H. E. Armstrong in his Huxley memorial lecture. He said that, 'Though highly skilled as a minute anatomist and as a close observer of structure, he was never an experimentalist but essentially a didactic teacher—with a marked tendency to pontificate. Had he been otherwise, more consciously formative and persuasive as a teacher, his influence would have been far greater: he would have secured more than the select following of intimates.'

How, then, could Drummond maintain such an apparently one-sided view? He was writing in the 1890s, a period of scientific toleration in which men could look back with complacency on gigantic strides in science, physical and biological, pure and applied, over the previous half century, and where there was still room for a Whiggish optimism. It is interesting to compare the sentiments in Drummond's quotation with the following extract from a speech by the German chemist Kekulé, delivered in 1890 and looking back to events twenty or thirty years before. The subject is different—chemistry not biology—and the details do not matter, but an underlying similarity of emphasis is apparent. Note in each case the sudden flash of enlightenment.

> Fifty years ago, the stream of chemical progress had divided into two branches . . . at length, as the two branches had again approached much nearer to one another, they were separated by a thick growth of misunderstandings, so that those who were sailing along on the one side neither saw those on the other, nor understood their speech. Suddenly a loud shout of triumph resounded from the host of the adherents of the type theory. The others had arrived, Frankland at the head. Both sides saw that they had been striving towards the same goal, although by different routes. They exchanged experiences; each side profited by the conquests of the other; and with united forces they sailed onward on the reunited stream.

It appears to be probable, then, that although the Darwinian hypothesis did accomplish some degree of integration within science this was, not surprisingly, largely confined to the biological sciences and then to a rather more limited extent than might be inferred from the rapturous eulogies of Drummond.

4.2 The physical sciences

If as we have suggested the physical sciences were exempt, even on Drummond's account, from the integrative influence of Darwinism it must not be supposed that there was no intellectual commerce between physical and biological science in the nineteenth century. One single example is the new awareness of the possibility that the chemical elements had evolved in some way, a phenomenon which William Crookes proposed to call 'inorganic Darwinism.' Yet it cannot be denied that there was a considerable barrier between the biological and physical sciences at this time, a barrier that was in no way removed by the presence of Darwinism. Darwin himself, as his letters show, was gravely disturbed by the apparently uncomprehending attitudes to his work of mathematicians and physicists. His use of such phrases as 'we may suppose' (about 800 times, someone counted) was unfavourably contrasted with the assured language of the physical scientists. It has even been suggested by R. E. D. Clark that:

> Psychologically, the law of evolution by natural selection forwarded a means by which Darwin and his non-mathematical contemporaries could transform their feelings of inferiority into grounds of superiority. It enabled them to hope that henceforth real

scientific advance would be achieved without a grounding in physical and mathematical thought.

However that may be, it remains a fact that a great many of the leading figures in late Victorian physical science found the case for evolution unconvincing. Rarely does this seem to have arisen on religious grounds although, as we shall see, physical scientists of that period tended to view the claims of religion in a sympathetic way. The biological scepticism of Lord Rayleigh is typical. As late as 1906 he was asked by his son whether he could accept natural selection as a sufficient explanation of evolution and replied, 'Well, no, I don't think I can quite swallow it'.

It is hard to identify a clear reason for the physical scientists' conservatism with regard to evolution. Goodfield and Toulmin have suggested that 'Darwin's theory of organic evolution woke no echoes in the physical world, for there the tide of atomic theory was flowing strongly in the opposite direction', arguing that fixity of species, now denied in biology, was just becoming accepted in chemistry. Presumably the reference is to Dalton's atomic theory announced in the first decade of the century although the non-interconvertibility of chemical elements was an item of belief very much older than that. Perhaps one basic reason lay in the nature of the evidence for evolution: it simply lacked the rigour and precision deemed necessary in physical science.

Nevertheless it seems that there was more to it than that. While it would be a considerable exaggeration to say that Victorian physical scientists tended to be incorrigibly religious, there are sufficient numbers of them who did identify themselves strongly with varying theistic positions to suggest that they were more in the mainstream tradition of a religion–science alliance than were some of their biological colleagues. Nor was their attitude to a scientific reductionism merely passive; as Dr D. M. Knight has observed, 'Some of the most strenuous opponents of a wholly mechanistic biology have been physical scientists'. Amongst those who most obviously come into the category of physical scientists with clearly defined Christian beliefs are Brewster, Joule, Clerk, Maxwell, Kelvin, Rayleigh and Perkin. Kelvin, for example, wrote of the 'utter futility of his [Darwin's] philosophy' and in a British Association Presidential Address (1871) concluded:

> I feel profoundly convinced that the argument of design has been greatly too much lost sight of in recent zoological speculations. Reaction against the frivolities of teleology, such as are to be found, not rarely, in the notes of the learned commentators on Paley's 'Natural Theology,' has I believe had a temporary effect in turning attention from the solid and irrefragable argument so well put forward in that excellent old book. But overpoweringly strong proofs of intelligent and benevolent design lie all round us; and if ever perplexities, whether metaphysical or scientific, turn us away from them for a time, they come back upon us with irresistible force, showing to us through Nature the influence of a free will, and teaching us that all living beings depend on one ever-acting Creator and Ruler.

In somewhat similar vein Lord Rayleigh insisted on prefacing his collected papers with the text

> The works of the Lord are great, sought out of all them that have pleasure therein.

However, he had to report that 'the Secretary to the Press suggested with many apologies that the reader might suppose that *I* was the Lord'!

But in a sense the father of them all was Michael Faraday whose work has come to dominate science in a more complete way even than that of Darwin. Although he did not live to see the surge of Darwinism reach its floodtide (he died in 1867) Faraday's attitudes are worth a brief examination. They were far from untypical.

Faraday's achievements were immense and he has often, like Darwin, been accorded the ultimate compliment of comparison with Newton. In chemistry an amazing deftness in manipulation helped him to liquefy chlorine and isolate benzene, and, building on the foundations laid by his mentor Davy, he discovered his two laws of electrolysis and the phenomenon of electromagnetic induction, the basis of the dynamo. It was to Faraday, more than to anyone else, that physics was indebted for the elimination of the assortment of weightless fluids with which it was cluttered at the beginning of the last century: electricity (two kinds), caloric, magnetism (several varieties), light and others. Inspired by a deep conviction regarding the unity of natural forces he was led to adumbrate his Field Theory which, in the hands of Clerk Maxwell, led to a mathematical revolution in theoretical physics and ultimately to the prediction and discovery of radio waves.

Figure 14 Michael Faraday, painting by Sir Thomas Phillips, 1842 (National Portrait Gallery).

Now it has often been argued that Faraday kept his science and religion well apart. So, for example, Gillispie thought (*Gillispie*, p. 208). Recent work has suggested that this is far from the truth, however. Of Faraday's religion he himself once said:

> I am of a very small and despised sect of Christians, known, if known at all, as Sandemanians, and our hope is founded on the faith that is in Christ.

This minority noncomformist grouping, founded by Robert Sandeman in 1771, was ascetic in life-style, democratically non-clerical in organization, and evangelical in theology. The least important, but most curious, aspect of its practice was a revival of the New Testament practice of feet washing.

To help to decide whether Faraday's religion and science were as insulated from each other as Gillispie implied it would be helpful to examine a selection of Faraday's remarks. The quotations below are all his; please read them in the light of the questions which follow.

> A
> The Christian finds his guide in the words of God and commits the keeping of his soul into the hands of God. He looks for no assurance beyond what the Word can give him, and if his mind is troubled by the cares and fears which assail him he can go nowhere but in prayer to the throne of grace and to Scripture.
>
> B
> The Christian religion is a revelation. The natural man cannot know it.
>
> C
> Let no-one suppose for a moment that the self-education I am about to commend, in respect of the things of this life, extends to any considerations of the hope set before us, as if man by reasoning could find out God. It will be improper here to enter upon this subject further than to claim an absolute distinction between religious and ordinary belief. I shall be reproached with the weakness of refusing to apply those mental operations which I think good in respect of high things to the very highest. I am content to bear the reproach; yet even in earthly matters, I believe that the invisible things of Him from the creation of the world are clearly seen, being understood by the things that are made, even His eternal power and Godhead; and I have never seen anything incompatible between those things of man which are in him, and those higher things concerning his future, which he cannot know by that spirit.
>
> D
> Though the natural works of God can never by any possibility come in contradiction with the higher things that belong to our future existence, and must with everything concerning Him ever glorify Him, still I do not think it at all necessary to tie the study of natural science and religion together, and in my intercourse with fellow creatures, that which is religious, and that which is philosophical, have ever been two distinct things.
>
> E
> Is the lingering notion, which remains in the minds of some, really a thought, that God could not just as easily by His word speak power into existence around centres, as He could create nuclei and then clothe them with power?

1 What do these quotations tell us of Faraday's theology of nature and his natural theology?
2 Which of these seems out of key with all the others?

To deal with the second question first, it must surely be clear that the odd one is the last, for this alone deals with a theology of nature; in other words it is only in this one that theological considerations are applied to topics within science. It comes from an unpublished manuscript note dated 18 February 1844 now in the library of the Institution of Electrical Engineers. All the other quotations in effect deny a natural theology in the sense in which it was used in the eighteenth century. Like many other evangelicals Faraday became aware of the limitations of natural theology and the primacy of revelation in the convincing of a man. Each of the first four extracts really says just this and emphasizes the importance he placed upon Scriptural revelation. Perhaps the nearest Faraday got to anything like the classical physico-theology was in a lecture on putrefaction delivered in 1840 when he spoke of 'evidence of a wisdom which the more a man knows the more freely will he acknowledge that he cannot understand', but that was not going very far. It is because of his persistent underrating (by conventional standards) of natural theology[1] that the impression has been created that he kept religion and science in two separate compartments. But the last quotation decisively refutes that position. It was because of his conception of God that he adopted a certain view of atomic constitution from which, indeed, emerged many fruitful speculations about the nature of forces and which led him, ultimately, to his Field Theory. So although he claimed, quite rightly, that 'there is no philosophy [science] in my religion' the converse was certainly not true. In this respect he was entirely typical of many other physical scientists of his century as, indeed, of some earlier periods. But it is interesting to see that natural theology received setbacks not merely within the field of evolutionary biology but also from some of the most strenuous opponents of that philosophy as indeed, it also did from the theologians like Karl Barth (see p. 57).

Figure 15 Sir William Thomson, later Lord Kelvin, photograph by W. and D. Downey (Mansell Collection).

The religious commitment of many of the most prominent physical scientists in the nineteenth century may not have increased their predilection for evolution, but that is only part of the story. It was from their scientific researches that objections were sometimes raised that proved difficult or even insurmountable within Darwin's own lifetime. Doubtless the effect of this was to reinforce their earlier commitments and to deepen the chasm between them and those evolutionary biologists who chose to oppose the Christian religion. One of the chief protagonists was Sir William Thomson, later Lord Kelvin; the challenge came from the new science of thermodynamics, concerned with the study of heat and energy changes. Thomson did not oppose evolution as such, admitting it to be consistent with divine benevolence, but he was concerned with the time scale required. In 1862 he began to apply to both the sun and the earth laws relating to loss of heat. He concluded that the earth was cooling down at such a rate that only a few million years ago it would have been too hot to sustain life. The almost limitless ages demanded by the uniformitarian geologists and the evolutionary biologists had simply not been available. Darwin himself concluded, 'this objection . . . is one of the gravest as yet advanced'. The fact that Thomson, in the light of twentieth-century knowledge, was proved to be wrong is not really the point. His attack simply illustrates that, even on purely scientific grounds, no side had it all its own way. What he did not know was that within both earth and sun lie unimaginable potential for the development of energy in phenomena of nuclear reactions: either spontaneous disintegration, as in the terrestrial radioactivity discovered at the end of the century, or in nuclear fusion recognized in the sun in our own day. The former develops sufficient heat to slow down the rate of cooling of the earth. It is now perfectly possible to suppose that the earth is old enough to have supported biological evolution at the kind of rate originally envisaged.

Thermodynamics furnished further assistance in the physicists' harrassment of evolution in another way. The second law of thermodynamics implies a running down of the universe—and of any closed system—as its entropy, or degree of disorder, increases. An example of this mode of attack appeared in *The Unseen Universe*, 1875, by two Scottish physicists, Arthur Stewart and P. G. Tait. The book is much in the Faraday tradition; indeed Faraday and Newton are cited as examples of the compatibility between science and religion. It was intended as a reply to an address by

[1] In this, of course, he differed from Kelvin.

Tyndall of 1874 which proposed a materialistic conception of the universe partly on the basis of the law of conservation of energy (the first law of thermodynamics), regarded as a testimony to the universal reign of scientific law so essential to a materialist position. Stewart and Tait assailed Tyndall on the effective grounds that he had emphasized the first law and conveniently forgotten the second. This led them to a number of technical considerations which appeared to invalidate Tyndall's thesis. A few years later Haeckel made almost exactly the same mistake in arguing on the basis of the first law only. Since then other writers have seen in this second law a fundamental objection to the basic concepts of evolutionary biology. However, that is another story and it is wise to regard the matter as still open.

4.3 Some social aspects

Finally it remains to touch on the theme of science in its social context, and, in particular, the way in which this was affected by the debates involving science and religion.

Perhaps there was no more important or potentially serious threat to science than that associated with its 'public image' in the later half of the nineteenth century. At all times this was to be closely linked with the educational enterprise of Victorian Britain. Prior to the publication of the *Origin* there had been considerable opposition to the introduction of science into some areas of elementary instruction on the grounds that it would displace the teaching of religion. More serious, however, was the surge of public opinion against science in general in the last quarter of the century. The Victorian 'swing from science' was undoubtedly connected with ideological attitudes, and many saw an immediate cause in Tyndall's 1874 address which, in the scientific fold, brought the swift reaction from Stewart and Tait.

Tyndall's plea for 'materialism' was reinforced by the continued pressures exerted by Huxley who sought to detach science from organized religion with such vigour that his agnosticism undoubtedly induced an over-reaction. As David Layton has recently pointed out, 'an awareness of religious misgivings about the natural history sciences would seem to have encouraged the Jermyn Street practitioners[1] to over-emphasize the respectability of their studies'.

During the 1880s there were repeated attacks on science from literary critics, and not only those with a religious axe to grind. A genuine fear for a society threatened by absorption into the jungle was coupled with a profound distaste for reductionist views of man. Undoubtedly the polarization of views sought by Huxley was achieved, but at considerable cost to a science that was struggling to gain public approval and (in many cases) financial support. One is reminded of the advice given by Turgenev: 'If you're going to kick down the pillars of society don't forget your own legs'.

Perhaps it was inevitable that there should be numerous groupings and regroupings amongst scientists of that period. One of the most powerful unofficial groups to emerge was the so-called 'X-club'. The X-club was an association of nine scientists in mid-Victorian days who met as an unofficial body and pressure-group. The membership consisted of the following: Busk, Frankland, Hirst, Hooker, Huxley, Lubbock, Spencer, Spottiswood and Tyndall. It was said by Frankland that 'All these colleagues of mine occupied some of the highest positions in the scientific world, and were of one mind on theological topics.' A glance at the list should not leave one in any doubt as to the nature of their singlemindedness. Indeed it has been urged recently that one of their prime motivations was a desire to ward off attacks on science by conservative theologians, and their private journals reveal that this was true. They sought to meet 'pure and free, untrammelled by religious dogmas'. The X-club was perhaps unusual in its importance and influence but there can be no doubt that regular meetings reinforced

[1] Jermyn Street was the home of the Museum of Practical Geology and the Government School of Minds consolidated in the new Department of Science and Arts.

the group's solidarity in matters of religious agnosticism. The X-club was a relatively permanent institution meeting for more than twenty years from 1864.

Another group, however, was convened in 1865 for precisely the opposite reason. This was the Victoria Institute or Philosophical Society of Great Britain whose first President was the Earl of Shaftesbury, succeeded by Sir George Stokes in 1886. The objects were to promote investigations into the relationship between science and Christian faith. In its early years it was a forum for some interesting debates in which scientists and clerics joined in amity. It continues to function today. Interestingly, Clerk Maxwell declined to join, not on the grounds that a synthesis between science and religion was undesirable or even impossible, but simply that it was too personal a matter for public debate.

In the same year a further event took place at the British Association meeting: the issue of a 'Declaration of Students of the Natural and Physical Sciences'. Typical of the Victorian penchant for signed petitions it acquired 617 signatures many of which, unlike the names of the X-club, are relatively unknown. But that is perhaps just the point. They represent a genuine reaction on the part of ordinary students or practitioners of science against the excesses of those who claimed that science had finally rendered the authority of Scripture untenable. The Declaration, which now rests in the Bodleian Library at Oxford, is reproduced below. As an expression of popular opinion at the height of the Darwinian controversy it is not an inappropriate document with which to conclude this course.

The Declaration of Students of the Natural and Physical Sciences (1865)

We, the undersigned Students of the Natural Sciences, desire to express our sincere regret, that researches into scientific truth are perverted by some in our own times into occasion for casting doubt upon the Truth and Authenticity of the Holy Scriptures. We conceive that it is impossible for the Word of God, as written in the book of nature, and God's Word written in Holy Scripture, to contradict one another, however much they may appear to differ. We are not forgetful that Physical Science is not complete, but is only in a condition of progress, and that at present our finite reason enables us only to see as through a glass darkly; and we confidently believe that a time will come when the two records will be seen to agree in every particular. We cannot but deplore that Natural Science should be looked upon with suspicion by many who do not make a study of it, merely on account of the unadvised manner in which some are placing it in opposition to Holy Writ. We believe that it is the duty of every Scientific Student to investigate nature simply for the purpose of elucidating truth, and that if he finds that some of his results appear to be in contradiction to the Written Word, or rather to his own *interpretations* of it, which may be erroneous, he should not presumptuously affirm that his own conclusions must be right, and the statements of Scripture wrong; rather, leave the two side by side till it shall please God to allow us to see the manner in which they may be reconciled; and, instead of insisting upon the seeming differences between Science and the Scriptures, it would be as well to rest in faith upon the points in which they agree.

1 In this Declaration what similarities are to be found to sentiments expressed in earlier times?
2 What new features are evident?
3 What basic presuppositions underlie the Declaration?

The main sentiments in this document must be surely familiar to all who have followed the course to its end. Time and again there have been signs of an apparent conflict between religion and science, and time and again this has been shown (or claimed) to have originated in a failure to recognize two things: the incompleteness of scientific knowledge and the vulnerability of individual interpretations of Scripture, particularly in matters relating to natural science. The reference to the two books, of Nature and Scripture, goes back to long before the eighteenth century (see Unit 2, p. 69). The plea to suspend judgement on disputed matters and the faith in an ultimate reconciliation between science and Scripture could well have come from the earlier proponents of 'accommodation', such as Galileo (*Anthology I*, p. 27), Calvin (*Hooykaas*, p. 120, especially the middle paragraph, and p. 122) and even Augustine.

The statement is so thoroughly in line with earlier writings that it is hard to detect real elements of novelty. One exception perhaps is the suspicion now cast upon natural science. In previous crises sections of the church had taken a low view of science, but it was in Victoria's reign that a disenchantment became widespread in society (compare above, pp. 65–6). The signatories were also aware that science was being used by some for ulterior purposes. Doubtless to them this *was* a new and unwelcome departure.

Behind this Declaration lie several assumptions which a philosopher might question. There is a belief in the scientific enterprise as a means of elucidating truth, a conception of nature as in some way a manifestation from and about God, and a confidence that much fuller knowledge will one day be available. But underlying all is an apparently unshakeable conviction concerning the nature of Scripture. Traditionally this had been universally regarded by the church as the Word of God, and now, despite the changes in theological opinion described in Unit 14, the signatories reaffirm the same belief. However, what is remarkable about this document, considering its unsophisticated origin, was not its conservativeness about Scripture but rather the opposite. Although they fully accepted the truth of the Bible, they were careful to distinguish this from their own interpretation and to leave wide open the exact nature of that truth.

Thus as we end this course it is apparent that though much had changed, many of the principal themes recur again and again: the problem of design, the religious incentive to scientific research, the importance of other extraneous considerations on decision making in science, the grouping and re-grouping of scientific workers, the compulsion to look at life whole and the impossibility of insulating science from other belief systems—all these and many more themes have characterized the progress of science down the centuries. But underlying so much was this question of Scripture and changing attitudes to it. It was, after all, the Bible which posed the problems to which 'accommodation' was a suggested answer, and it was upon the Bible that all Christian theology, Protestant and Catholic, was theoretically constructed. The notion that the essence of the problem of science and belief was a conflict between enlightened science and Biblical obscurantism is just too naïve and too crude to survive serious historical enquiry. But despite this there can be no denying that the Scriptures lay at the heart of the problem. At first science progressed because of, not in spite of, the Bible; later its dependence was far less obvious but no less real, as Judaeo-Christian ideals continued to act as a mainspring for the belief in scientific progress, the social utility of sciences, the worthwhileness of the whole enterprise and the ideological motivation of not a few of its pioneers. It was in the changing interactions between science and Scripture that man's awareness of nature and of himself became most acutely realized. Perhaps it still is.

Figure 16(a) The great doors of the Cavendish Laboratory, Cambridge, one of the world's leading centres for scientific research since its foundation in 1871 (*Photo Edward Leigh*).

Figure 16(b) The words carved on the doors of the Cavendish Laboratory, epitomizing the religious approach to science characteristic of many physical scientists in the late Victorian era. They read: '*MAGNA OPERA DOMINI EXQUISITA IN OMNES VOLUPTATES EIUS*', '*The works of the Lord are great, sought out of all them that have pleasure therein*' (*Photo Edward Leigh*).

Further Reading

On the reception of Darwinism see the bibliographies in earlier units (especially 14) and in *Gillispie*. Also useful is R. E. D. Clark, *Darwin: Before and After*, Paternoster Press, London, 1966. The impact of Darwinism on the church is further discussed in A. R. Vidler, *The Church in an Age of Revolution*, Pelican, 1961 (especially chapter 10) and Alan Richardson, *The Bible in the Age of Science*, S. C. M. Press, London, 1964, and in John Dillenberger, *Protestant Thought and Natural Science*, Collins, London, 1961. Relevant papers occur in the *British Journal for the History of Science:* e.g. R. Smith, 'Alfred Russel Wallace: Philosophy of Nature and Man', 1972, **6**, 177–99, and M. Bartholomew: 'Lyell and Evolution: an account of Lyell's response to the prospect of an Evolutionary Ancestry for Man', 1973, **6**, 261–303. The same journal has some interesting articles on religious attitudes in physical science, e.g. T. H. Levere, 'Faraday, Matter and Natural Theology—Reflections on an Unpublished Manuscript', 1968, **4**, 95–107, and P. M. Heimann, '*The Unseen Universe:* Physics and the Philosophy of Nature in Victorian Britain', 1972, **6**, 73–9.

The literature on the organization of science in England is vast, and many writers have touched on the effects of religious doubt on the sociology of science, e.g. David Layton, *Science for the People*, Allen and Unwin, London, 1973, and R. M. MacCleod, 'The Support of Victorian Science: The endowment of Research Movement in Great Britain, 1868–1900', *Minerva*, 1971, **4**, 197–230. On the X-club see J. V. Jensen, 'The X-club: Fraternity of Victorian Scientists', *British Journal for the History of Science*, 1950, **5**, 63, and R. M. MacCleod, 'The X-club, a Social Network of Science in Late Victorian England', *Notes and Records*, 1970, **24**, 181.

Finally, looking back over the whole course, the following general books may be recommended for the different insights they offer and, in most cases, for the excellence of their general surveys.

John Dillenberger, *Protestant Thought and Natural Science* (as above).

I. G. Barbour, *Issues in Science and Religion*, S. C. M. Press, London, 1966.

C. E. Raven, *Christianity and Science*, Lutterworth, London, 1956.

Alan Richardson, *The Bible in the Age of Science* (as above).

M. A. Jeeves, *The Scientific Enterprise and Christian Faith*, Tyndale Press, London, 1969.

Not all of these books concentrate upon the historical aspects of the subject nor do all pretend to be scholarly monographs, but each has provocative things to say about the issues that have been discussed and each is worth examination even if, at the end of it, you do not agree with the conclusion of the authors.

Acknowledgements

Grateful acknowledgement is made to the following sources for illustrations used in this unit:

Figures 1, 7, 9a, 9b, 10, 11, 12 and 15: Mansell Collection; *Figures 2, 16a and 16b:* Photos by Edward Leigh; *Figure 5:* Bettmann Archive; *Figures 6, 8 and 13:* Mary Evans Picture Library; *Figure 14:* National Portrait Gallery.

Science and Belief: from Copernicus to Darwin